From Machine-to-Machine to the Internet of Things

Introduction to a New Age of Intelligence

From Machine-to-Machine to the Internet of Things

Introduction to a New Age of Intelligence

Jan Höller

Vlasios Tsiatsis

Catherine Mulligan

Stamatis Karnouskos

Stefan Avesand

David Boyle

ELSEVIER

AMSTERDAM • BOSTON • HEIDELBERG • LONDON
NEW YORK • OXFORD • PARIS • SAN DIEGO
SAN FRANCISCO • SINGAPORE • SYDNEY • TOKYO

Academic Press is an imprint of Elsevier

Academic Press is an imprint of Elsevier
The Boulevard, Langford Lane, Kidlington, Oxford, OX5 1GB
225 Wyman Street, Waltham, MA 02451, USA

First published 2014

British Library Cataloguing-in-Publication Data
A catalogue record for this book is available from the British Library

Library of Congress Cataloging-in-Publication Data
A catalog record for this book is available from the Library of Congress

ISBN: 978-0-12-407684-6

For information on all Academic Press publications
visit our website at **store.elsevier.com**

Printed and bound in the UK

14 15 16 17   10 9 8 7 6 5 4 3 2

Contents

PART 1 THE VISION FOR MOVING FROM M2M TO IOT

Foreword

I grew up in a time when the Internet was used by computer science students using Gopher to browse their course syllabus. We ran private bulletin-board systems using ANSI text over 2400 baud modems over fixed phone lines, and transfered news and mailing lists overnight through USENET. Think of this as an analogy to where we have been with automation systems and M2M over the past decade. The same incredible growth of people using the Internet in the 1990s is now being repeated by things using the Internet in the 2010s.

It is wonderful to see this book published during the peak of the IoT hype cycle, where most writing is in tweets and blog entries. The deployment of traditional IP networks, security technology and Web infrastructure requires a lot of knowledge and skill, and understanding the Internet of Things requires a similar breadth of knowledge. Today we take that knowledge for granted because we have trained the world through books and teaching over several decades. Luckily most of the knowledge we have gained from building today's Internet and Web services can be applied to IoT. There are, however, many aspects of IoT technologies that are new, including IPv6 over low-power networks, new applications of TLS security, efficient web transfer protocols and techniques for managing and using devices through commonly understood data objects.

System and network architects, administrators and software developers will find this book useful as an overview of IoT architecture and technology. At the same time, business and product managers will find the book useful as an introduction to the market segments, applications and requirements as input for a successful IoT product or service. Finally, the technology overview is a great starting place to find the information needed to dive deeper into a particular area, and the architecture overview covers a wide range of design paradigms. One important point made is that without trust and security built into IoT technology and systems in a holistic way, we will not see an Internet of Things, but continue to see silos of things.

The technology is available today to build an Internet of Things where devices and services can be developed and deployed for the benefit is society and industry as a whole. The challenge now is for us to educate people.

Zach Shelby
Directory of Technology for IoT
ARM Inc.

Foreword

Taking notice of the IoT — Since the age of 9, when I started programming, I thought *computers* were cool. At 15, I was hired to hack networks, catching the attention of some newspapers, and thought *networks* were cool. As a Lieutenant in the Air Force at the Pentagon, I was helping build the Arpanet and still thought *networks* were cool. In 1996 while helping design IPv6, I wrote the first implementation of v6 for the PC and in 2001 rewrote it for an 8-bit micro-controller and realized that *embedded networks* were cool. Most recently helping found and grow the IP for Smart Objects Alliance and now serving as White House Presidential Innovation Fellow working on Cyber-Physical Systems and the Internet of Things, I'm seeing everyone take notice of how CPS, the IoT and M2M will reshape our world and that is *really cool*.

In 1999 Scott McNealy quipped, "You have zero privacy anyway... Get over it." We should <u>not</u> get over it, but instead deal with it. It is critical that we think about it — privacy — experiment with it and work to get out in front of the issues rather than play catch up. Books like this one are important in bringing the concepts and ideas related to this new emerging smarter world into focus for discussion and debate. According to a recent survey, the United States now has more Internet connected gadgets, sensors, controllers, phones and light blubs, than the 311 million people that live in the US. Understanding architectural design trade-offs with the application to specific implementation scenarios is important if we are to get this right.

It is fundamentally important that the Internet of Things and these Machine-to-Machine networks are built using open standard protocols — especially IP. Jari Arkko, the current IETF Chairman, started describing "permissionless innovation" whereby new businesses, new systems and new business models can be created without having to ask permission from others. Open protocols and open standards set the stage for these opportunities. When Vint Cerf and others created the Internet, they didn't plan for YouTube or Facebook, but their layered network design and freely available protocols allowed for these types of innovation.

There are quite a few books about the Internet of Things, but few of them provide an accessible description of a vision for the connected world and the basic building blocks necessary to bring this vision to reality. But this book goes beyond those fundamentals to sharing specific examples in Asset Management, Industrial Automation, the Smart Grid, Commercial

Building Automation, Smart Cities (a particular favorite since it aligns with my Presidential Innovation Fellow project — The SmartAmerica Challenge), and Participatory Sensing. Teasing apart the important nuances that differentiate each of the application spaces is critical in understanding how to apply sound design to each. Within each segment are differing requirements for latency, security, privacy, determinism, throughput and speed. Understanding these differences is critical for a proper system design and successful installation and deployment. This book can provide just such necessary information.

The next generation of devices that will become part of the Internet of Things will not just sense and report, but will control. Whether is it the connected vehicle, a building automation system, an agile manufacturing robot, a thermostat or a door lock, these new connected machines will have a greater impact on our lives. The protection of the control data and operating instructions will be critical as we allow greater control and autonomy so as to ensure our safety and security. Privacy and security "by design" is imperative and must not be an afterthought.

As we rush toward 2020 and the 50 billion Internet of Things as predicted by Ericsson, we need to be thoughtful, clue-full and have a plan so as not to be crushed by the onslaught of device management, privacy concerns and the avalanche of data. It's been over a decade since Kevin Ashton first used the term the Internet of Things. Progress has been slowed by the deployment of islands of proprietary protocols and the gateway required to interconnected them; the proliferation of pseudo-open standard (but yet proprietary) protocols and yet more gateways required for interconnection and the continued quest for new and "better" protocols. We have the necessary tools at hand. It is the application of sound design and open standards that will allow us to march into this new era of the connected everything with a confidence that it holds the promise of a safer and more efficient world and society — it will be *awesomely cool*.

Geoff Mulligan
Presidential Innovation Fellow
Founder, IPSO Alliance

Acknowledgements

A work of this nature is not possible without other's support and input. The authors would like to gratefully acknowledge the contribution of many of our colleagues at Ericsson, SAP and Imperial College, London, as well as our colleagues across industry and academia.

We would like to acknowledge many of colleagues at SAP for the fruitful discussions, in particular: Domnic Savio, Patrik Spiess, Dejan Ilic, Per Goncalves Da Silva, Luciana Moreira Sá de Souza, Dominique Guinard, Vlad Trifa, Oliver Baecker, Moritz Köhler, Nina Oertel, Zoltan Nochta, Stephan Haller, Anke Weidlich, Harald Vogt and Orestis Terzidis.

We would also like to acknowledge our colleagues at Ericsson for many good discussions and support, and in particular: Srdjan Krco, Jari Arkko, Bo Svensson, John Fornehed, Magnus Olsson, Ioannis Fikouras, Andras Toth, Andreas Fasbender, Göran Selander, Mattias Eld, Rickard Cöster, Konstantinos Vandikas, Vincent Huang, Sebastien Pierrel, Elena Fersman and Niklas Björk.

In addition many of the views depicted here have been the result of discussions over years with several people from industry and academia. We would especially like to mention in alphabetical order: Thomas Bangeman, Martin Bauer, Tim Bauge, Jesús Bernat Vercher, François Carrez, Fabien Castanier, Armando Walter Colombo, Jerker Delsing, Deborah Estrin, Roch H Glitho, Alex Gluhak, Richard Gold, François Jammes, Pedro José Marrón, Jose L. Martinez Lastra, Arne Munch-Ellingsen, Mirko Presser, Jochen Rode, Zach Shelby, Petr Stluka, Rikard Strid, Mani Srivastava, Marco Taisch, and Joachim Walenski.

We would also like to thank our families. Writing this book would not have been possible without their generosity and support throughout this process.

Dr. Catherine Mulligan and Dr. David Boyle would like to acknowledge the funding provided by the RCUK Digital Economy Program, specifically Sustainable Society Network + (EP/K003593/1) and Digital City Exchange (EP/I038837/1). In addition, they would like to thank their colleagues on these projects for fruitful discussions.

We would like to also acknowledge the following European Commission funded projects and their partners: IMC-AESOP (www.aesop.eu), SOCRADES (www.socrades.eu), SmartKYE (www.SmartKYE.eu),

NOBEL (www.ict-nobel.eu), SmartHouse/SmartGrid (www.smarthouse-smartgrid.eu), makeSense (www.project-makesense.eu), CoBIs (www.cobis-online.de), SENSEI, IoT-A (www.iot-a.eu), IoT-i (www.iot-i.eu) and CONET (www.cooperating-objects.eu).

Author Biographies

Jan Höller is a Principal Researcher at Ericsson Research where he has a responsibility to define and drive technology and research strategies, and to contribute to the company strategies in the area of M2M and Internet of Things. He established Ericsson's research activities in the Internet of Things almost a decade ago, and has since continued to contribute to the company strategies in the area of M2M and Internet of Things towards the Ericsson vision of "50 Billion connected devices" in the Networked Society. Jan has held various positions in Strategic Product Management, Technology Management and has, since he joined Ericsson Research in 1999, led different research activities and research groups. He also serves as secretary on the Board of Directors at the IPSO Alliance.

Vlasios Tsiatsis is a Senior Researcher at Ericsson Research, Ericsson AB. He holds a Ph.D. in the area of Networked Embedded Systems from the University of California, Los Angeles focusing on energy management of Wireless Sensor Networks. At Ericsson Research, he applied his sensor network expertise on IoT-related European Union projects such as RUNES, SENSEI, IoT-i and CityPulse as well as internal Ericsson corporate research projects around ma-chine/man/mobile-to-machine and IoT services. Vlasios has extensive theoretical and practical experience on IoT technologies and deployments and his research interests include system architecture, management of complex and heterogeneous systems including IoT, semantic technologies and their application on IoT systems as well the management of data emanating from large IoT deployments.

Dr. Catherine Mulligan is a Research Fellow in the Innovation and Entrepreneurship group at Imperal College, London. Catherine has extensive experience in the role of IoT in smart cities and global supply chains. She led the Urban Prototyping (UP) Festival, London — a six month festival dedicated to implementing and creating a flourishing IoT ecosystem in London. Catherine is Principal Investigator on two RCUK Digital Economy Program grants: Sustainable Society Network + and Scaling the Rural Enterprise. In addition, Catherine is a Co-Investigator on the EPSRC funded "Unleashing the Value of Big Data" and the Cloud Intelligent Protection At RUN-TIME (CIPART) projects and a Researcher on the Digital Cities Exchange project at Imperial College, London. Catherine has 15 years international experience in the Mobile Telecommunications and ICT industries, including 10 years at Ericsson in Stockholm, Sweden.

Working on a variety of cutting edge technologies, Catherine experienced first-hand the complexities of successfully taking innovation to market. Her research interests lie in the area of new economic and business models enabled by the digital economy. In particular, Catherine is interested in the role that technologies play in the creation of citizen-centric smart/sustainable cities.

Stamatis Karnouskos is an Expert on M2M / Internet of Things within SAP. He investigates the added-value of integrating networked embedded devices in enterprise systems. For more than 15 years, Stamatis has led efforts in several European Commission and industry funded projects related to industrial automation, smart grids, Internet-based services and architectures, software agents, mobile commerce, security and mobility. He serves in the technical advisory board of Internet Protocol for Smart Objects Alliance (IPSO), and the Permanent Stakeholder Group of the European Network and Information Security Agency (ENISA).

Stefan Avesand is a Senior Software Researcher at Ericsson AB where he drives research within the domain of Artificial Intelligence, and how to apply this in network operations and smart devices (such as in predictive maintenance and the Ericsson Social Web of Things concept). Stefan has 15 years of experience from the field of telecommunications, both from a research and a product development perspective, as well as from an operator and a manufacturer perspective. He has been active in the Internet of Things area for several years and was recently managing the development of an operator based Smart Home solution at Ericsson Research, where he coordinated with Strategic Product Management and external partners to provide services such as TV, multimedia, energy management and home security.

David Boyle is a Research Fellow in the Department of Electrical and Electronic Engineering at Imperial College, London. A member of the Optical and Semiconductor Devices Group, and contributing to the Digital Economy Laboratory, his research interests lie at the intersection of applied complex sensing, actuation and control systems (cyber-physical systems), data analytics, and digital economy. David received his PhD in Electronic and Computer Engineering from the University of Limerick, Ireland, in 2009, having graduated with a B. Eng. (Hons) in Computer Engineering in 2005. His work has been recognized and awarded internationally, and published in the leading technical journals, including the IEEE Transactions on Industrial Electronics (TIE), and Informatics (TII). He actively participates in a number of Technical Program and Organizing Committees, notably Design,

Automation and Test in Europe (DATE), and the ACM Conference on Embedded Networked Sensor Systems; the premier conference in the field. Before joining Imperial, David worked with Wireless Sensor Network and Microelectronics Applications Integration Groups in the Microsystems Centre at Tyndall National Institute, and the Embedded Systems Research Group, University College Cork, Ireland, primarily developing "green" wireless sensor networks to enable sustainable structural health monitoring. Previously, he was with France Telecom R&D — Orange Labs, focusing on end-to-end quality of service for urban machine-to-machine (M2M) services, and a Visiting Postdoctoral Scholar at the Higher Technical School of Telecommunications Engineering, Technical University of Madrid (ETSIT UPM).

The Vision for Moving from M2M to IoT

Part I of this book provides an overview of the vision and market conditions for M2M and the move towards IoT. Here we discuss the global context within which M2M and IoT exist, and the business and technical drivers at work in both technology and industry. This part provides the basics of the M2M/IoT-Architecture and the principles behind them, preparing the reader for Part II, which outlines in detail the architecture for M2M and IoT.

Introduction and Book Structure

CHAPTER OUTLINE

1.1 Introduction

This book provides a thorough and high-level analysis for anyone wishing to learn about how Machine-to-Machine (M2M) and the Internet of Things (IoT) are being implemented and deployed in various industries, and also cities. This chapter provides a brief introduction to the topics covered and the structure of the book.

The number of "connected devices" (i.e. devices connected to the Internet) is growing and is expected to continue to grow exponentially as people

From Machine-to-Machine to the Internet of Things: Introduction to a New Age of Intelligence.
DOI: http://dx.doi.org/10.1016/B978-0-12-407684-6.00001-2

increase the numbers of devices they purchase. Worldwide, mobile phone subscriptions have already exceeded 3 billion. End-users are also starting to use multiple devices (e.g. iPads, Kindles, mobile handsets, digital TVs, etc.). In addition, however, millions of new types of devices are emerging that allow machines to be connected to one another. These devices will communicate and offer services via the Internet, creating a new wave of innovation from both a technical and societal perspective. This explosive growth is unprecedented within not just the communications industries, but also the wider global economy.

This growth in the use of connected devices, M2M, and the IoT is expected to rapidly disrupt several business sectors in the next 5–10 years (Figure 1.1).

In addition, the traffic generated for M2M devices is predicted to grow 22-fold from 2011–2016.

In addition to all this, M2M solutions and services have a wider role to play in the future of our world. The year 2007 was a landmark year for the world: for the first time in history, more than 50% of the world's population was living in cities rather than rural areas (UN-HABITAT, 2011). This trend sees no signs of reversing. The infrastructure of cities and nations must therefore adapt accordingly, from roads, lighting, metro/commuter trains, and pipelines, to name just a few (HM Treasury 2011). Much of this infrastructure will be instrumented with sensors and actuators for more efficient management, and all these devices associated with infrastructure will be connected to large-scale data analysis and management systems, the data of which needs effective capture, analysis, and visualization in order to be applied effectively in the development of smart, sustainable societies and cities. In the UK alone, this market represents a significant investment by both the government and private sector alike. The use of M2M and IoT in assisting the delivery of economic, social, and environmental outcomes for nations and regions is rapidly becoming an area of concern for professionals working in this space (Broadband Commission 2012).

The unprecedented numbers of devices foreseen, in combination with the vertical nature of many M2M applications, create an interesting set of barriers to success for anyone wishing to implement a solution based on these technologies. The deployment and operation costs of traditional telecom platforms adapted to handle the traffic load from tens of billions of additional connected devices would be a prohibitively high investment. Moreover, due to the specialized nature of the cases where M2M technologies will be applied, a fragmented ecosystem is emerging in each of the

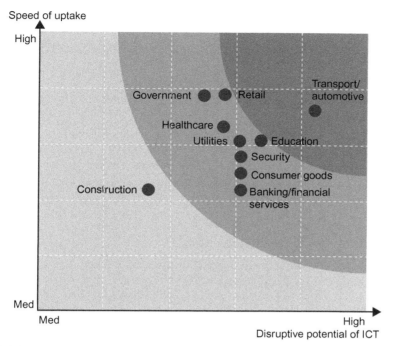

FIGURE 1.1

M2M disrupts several business sectors.

(Ericsson 2012)

solution "silos." Such industrial dynamics create barriers to entry for individuals and companies wanting to develop M2M applications or services, from supporting a mix of diverse devices and billing, to handling settlement and commission across the value chain.

Understanding how corporations and governments should respond to these changes is therefore a critical need for corporations, cities, and governments. The following section outlines the structure of the book as it covers these issues.

1.2 Structure of the book
Part I: The Internet of Things global context

Part I outlines the global context of M2M and the move towards IoT, including technology and business drivers.

Chapter 2: M2M to IoT — the vision
Chapter 2 provides an overview of how M2M solutions will move towards IoT, including the reasons that this is now occurring on the market.

Chapter 3: M2M to IoT — a market perspective
Chapter 3 provides an overview of the market drivers and industrial structures for the move from M2M to IoT.

Chapter 4: M2M to IoT — an architectural overview
Chapter 4 provides an overview of the architecture for IoT, including the overall design principles that sit behind the various architectures put forward by different standards bodies.

Part II: Nuts and bolts of M2M and IoT
In Part II, the technology building blocks of M2M and IoT solutions are presented, as well as the architecture.

Chapter 5: M2M and IoT technology fundamentals
This chapter provides an overview of relevant technologies for building M2M service solutions with a focus on the technologies that can be deployed widely, including: Devices and Device Gateways, Data Management, Business Process Engineering, and Cloud Technologies.

Chapters 6, 7 & 8: IoT architecture
These chapters together investigate how the different technologies introduced in Chapter 5 fit together in overall architectures, with reference to relevant system-level standards, including ETSI M2M and IoT-A.

Chapter 9: Real-world design constraints
Chapter 9 outlines design constraints that need to be taken into account when developing real-world technical solutions.

Part III: Implementation examples
Part III covers real-world implementation examples of M2M and IoT solutions.

Chapter 10: Asset management
Chapter 10 discusses Asset Monitoring, which enables the remote tracking and management of inventory in the field. Typically such functionality involves the collection of the exact location and state of assets at regular intervals for the purposes of improving the business (e.g. preventing stockouts) or reducing risks (e.g. of getting lost).

Chapter 11: Industrial automation
Chapter 11 covers the emerging approach in industrial environments, which is to create system intelligence by a large population of intelligent, small, networked, embedded devices at a high level of granularity, as opposed to the traditional approach of focusing intelligence on a few large and monolithic applications within industrial solutions.

Chapter 12: The smart grid
Chapter 12 covers the Smart Grid, a revolution currently transforming the electricity system. Rapid advances in IT are increasingly being integrated in several infrastructure layers of the electricity grid and its associated operations. M2M interactions create new capabilities in the monitoring and management of the electricity grid, and the interaction between its stakeholders.

Chapter 13: Commercial building automation
Chapter 13 covers commercial buildings and the use of IoT. The purpose of a building automation systems is typically to reduce energy and maintenance costs, as well as to increase control, comfort, reliability, and ease of use for maintenance staff and tenants. M2M and IoT are starting to play an increasingly important role within Commercial Building Automation.

Chapter 14: Smart cities
Chapter 14 covers Smart Cities, an emerging and increasingly important field of application for IoT. This includes how sensors and associated IoT systems are being applied and linked to other paradigms (e.g. open data initiatives).

Chapter 15: Participatory sensing
Chapter 15 covers Participatory Sensing (PS), or Urban, Citizen, or People-Centric Sensing. This is a form of citizen engagement for the

purposes of capturing the city surrounding environment and daily life. This chapter covers a few examples of such scenarios.

Chapter 16: Conclusion and looking ahead
Chapter 16 provides a brief overview of the future for IoT.

Part IV: Appendices
Part IV lists the abbreviations and references for the book.

M2M to IoT – The Vision

2.1 Introduction

Our world is on the verge of an amazing transformation; one that will affect every person, town, company, and thing that forms the basis of our society and economy. In the same way that the Internet redefined how we communicate, work, and play, a new revolution is unfurling that will again challenge us to meet new business demands and embrace the opportunities of technical evolution. Old and new industries, cities, communities, and individuals alike will need to adapt, evolve, and help create the new patterns of engagement that our world desperately needs. In response to these issues, we are moving towards a new era of intelligence – one driven by rapidly growing technical capabilities.

From Machine-to-Machine to the Internet of Things: Introduction to a New Age of Intelligence.
DOI: http://dx.doi.org/10.1016/B978-0-12-407684-6.00002-4

M2M and the IoT are two of the technologies that form the basis of the new world that we will come to inhabit. Anything in the physical realm that is of interest to observe and control by people, businesses, or organizations will be connected and will offer services via the Internet. The physical entities can be of any nature, such as buildings, farmland, and natural resources like air, and even such personal real-world concepts as my favorite hiking route through the forest or my route to work. This book covers our world's transformation towards the IoT, and it is the authors' hope that we will inspire solutions and provide a technical framework to some of the world's most pressing needs — from environmental change to industrial re-configuration.

2.2 From M2M to IoT

Within this book, we attempt to describe the move from what is today referred to as *Machine-to-Machine* communication towards an emerging paradigm known as the *Internet of Things*. Quite often these terms are used interchangeably. This chapter provides definitions of what we mean by the two terms and sets the stage for the rest of the book by outlining the global trends, capabilities, and drivers towards IoT.

2.2.1 A brief background

Both M2M and IoT are results of the technological progress over the last decades, including not just the decreasing costs of semiconductor components, but also the spectacular uptake of the Internet Protocol (IP) and the broad adoption of the Internet. The application opportunities for such solutions are limited only by our imaginations; however, the role that M2M and IoT will have in industry and broader society is just starting to emerge for a series of interacting and interlinked reasons.

The Internet has undoubtedly had a profound impact across society and industries over the past two decades. Starting off as ARPANET connecting remote computers together, the introduction of the TCP/IP protocol suite, and later the introduction of services like email and the World Wide Web (WWW), created a tremendous growth of usage and traffic. In conjunction with innovations that dramatically reduced the cost of semiconductor technologies and the subsequent extension of the Internet at a reasonable cost via mobile networks, billions of people and businesses are now connected to the Internet. Quite simply, no industry and no part of society have remained untouched by this technical revolution.

At the same time that the Internet has been evolving, another technology revolution has been unfolding — the use of sensors, electronic tags, and actuators to digitally identify, observe and control objects in the physical world. Rapidly decreasing costs of sensors and actuators have meant that where such components previously cost several Euros each, they are now a few cents. In addition, these devices, through increases in the computational capacity of the associated chipsets, are now able to communicate via fixed and mobile networks. As a result, they are able to communicate information about the physical world in near real-time across networks with high bandwidth at low relative cost.

So, while we have seen M2M solutions for quite some time, we are now entering a period of time where the uptake of both M2M and IoT solutions will increase dramatically. The reasons for this are three-fold:

1. An increased need for understanding the physical environment in its various forms, from industrial installations through to public spaces and consumer demands. These requirements are often driven by efficiency improvements, sustainability objectives, or improved health and safety (Singh 2012).
2. The improvement of technology and improved networking capabilities.
3. Reduced costs of components and the ability to more cheaply collect and analyze the data they produce.

What makes the M2M and IoT markets take off today, therefore, is *needs* meeting *enabling technologies* at the right *cost*.

The next section takes a closer look at the drivers of M2M towards IoT.

2.2.2 **M2M communication**

M2M refers to those solutions that allow communication between devices of the same type and a specific application, all via wired or wireless communication networks. M2M solutions allow end-users to capture data about events from assets, such as temperature or inventory levels. Typically, M2M is deployed to achieve productivity gains, reduce costs, and increase safety or security. M2M has been applied in many different scenarios, including the remote monitoring and control of enterprise assets, or to provide connectivity of remote machine-type devices. Remote monitoring and control has generally provided the incentive for industrial applications, whereas connectivity has been the focus in other enterprise scenarios such as connected vending machines or point-of-sales terminals for online credit card transactions. M2M solutions, however, do not generally allow for the broad sharing of data or connection of the devices in question directly to the Internet.

2.2.2.1 A typical M2M solution overview

A typical M2M system solution consists of M2M devices, communication networks that provide remote connectivity for the devices, service enablement and application logic, and integration of the M2M application into the business processes provided by an Information Technology (IT) system of the enterprise, as illustrated below in Figure 2.1.

The M2M system solution is used to remotely monitor and control enterprise assets of various kinds, and to integrate those assets into the business processes of the enterprise in question. The asset can be of a wide range of types (e.g. vehicle, freight container, building, or smart electricity meter), all depending on the enterprise.

The system components of an M2M solution are as follows:

- *M2M Device.* This is the M2M device attached to the asset of interest, and provides sensing and actuation capabilities. The M2M device is here generalized, as there are a number of different realizations of these devices, ranging from low-end sensor nodes to high-end complex devices with multimodal sensing capabilities.
- *Network.* The purpose of the network is to provide remote connectivity between the M2M device and the application-side servers. Many different network types can be used, and include both Wide Area Networks (WANs) and Local Area Networks (LANs), sometimes also referred to as Capillary Networks or M2M Area Networks. Examples of WANs are public cellular mobile networks, fixed private networks, or even satellite links.
- *M2M Service Enablement.* Within the generalized system solution outlined above, the concept of a separate service enablement component is also introduced. This component provides generic

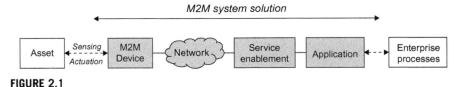

FIGURE 2.1

A generic M2M system solution.

functionality that is common across a number of different applications. Its primary purpose is to reduce cost for implementation and ease of application development. As we will see later and in Chapter 6, the emergence of service enablement as a separate system component is a clear trend.

- *M2M Application.* The application component of the solution is a realization of the highly specific monitor and control process. The application is further integrated into the overall business process system of the enterprise. The process of remotely monitoring and controlling assets can be of many different types, for instance, remote car diagnostics or electricity meter data management.

2.2.2.2 Key application areas

Existing M2M solutions cover numerous industry sectors and application scenarios. Various predictions have been made by analyst firms that provide market information such as key applications, value chains, and market actors, as well as market sizes (including forecasts) (ABI 2012, Berg 2013). A selected summary of main cellular M2M application markets is provided in Figure 2.2, and the figures are estimates of deployed numbers of corresponding M2M devices in the years 2012 and 2016, respectively.

The largest segment is currently **Telematics** for cars and vehicles. Typical applications include navigation, remote vehicle diagnostics,

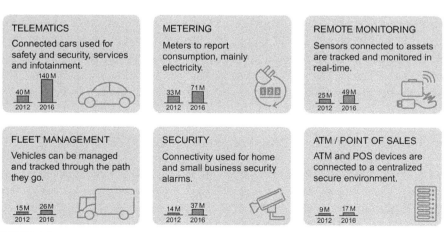

FIGURE 2.2

Summarized cellular M2M market situation.

(Courtesy of Ericsson)

pay-as-you-drive insurance schemes, road charging, and stolen vehicle recovery.

Metering applications, meanwhile, include primarily remote meter management and data collection for energy consumption in the electricity utility sector, but also for gas and water consumption.

Remote monitoring is more generalized monitoring of assets, and includes remote patient monitoring as one prime example.

Fleet management includes a number of different applications, like data logging, goods and vehicle positioning, and security of valuable or hazardous goods.

Security applications are mainly those related to home alarms and small business surveillance solutions. The final market segment is **Automated Teller Machines** (ATM) and **Point of Sales** (POS) terminals.

2.2.3 IoT

The IoT is a widely used term for a set of technologies, systems, and design principles associated with the emerging wave of Internet-connected things that are based on the physical environment. In many respects, it can initially look the same as M2M communication — connecting sensors and other devices to Information and Communication Technology (ICT) systems via wired or wireless networks.

In contrast to M2M, however, IoT also refers to the connection of such systems and sensors to the broader Internet, as well as the use of general Internet technologies. In the longer term, it is envisaged that an IoT ecosystem will emerge not dissimilar to today's Internet, allowing things and real world objects to connect, communicate, and interact with one another in the same way humans do via the web today. Increased understanding of the complexity of the systems in question, economies of scale, and methods for ensuring interoperability, in conjunction with key business drivers and governance structures across value chains, will create wide-scale adoption and deployment of IoT solutions. We cover this in more detail in Chapter 3.

No longer will the Internet be only about people, media, and content, but it will also include all real-world assets as intelligent creatures exchanging information, interacting with people, supporting business processes of enterprises, and creating knowledge (Figure 2.3). The IoT is not a new Internet, it is an extension to the existing Internet.

IoT is about the technology, the remote monitoring, and control, and also about where these technologies are applied. IoT can have a focus on the open innovative promises of the technologies at play, and also on advanced and complex processing inside very confined and close environments such as

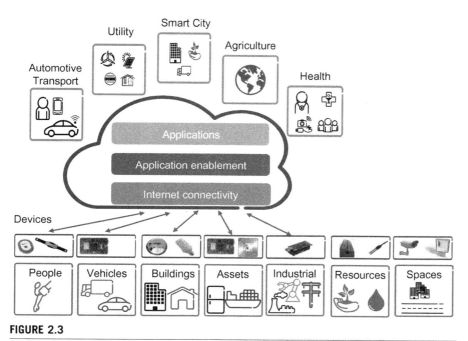

FIGURE 2.3

An IoT.

industrial automation. When employing IoT technologies in more closed environments, an alternative interpretation of IoT could then be "Intranet of Things."

Visions put forward (e.g. SENSEI 2013) have included notions like a global open fabric of sensor and actuator services that integrate numerous Wireless Sensor Network (WSN) deployments and provide different levels of aggregated sensor and actuator services in an open manner for application innovation and for use in not only pure monitor and control type of applications, but also to augment or enrich other types of services with contextual information. IoT applications will not only rely on data and services from sensor and actuators alone. Equally important is the blend-in of other information sources that have relevance from the viewpoint of the physical world. These can be data from Geographic Information Systems (GIS) like road databases and weather forecasting systems, and can be of both a static nature and real-time nature. Even information extracted from social media like Twitter feeds or Facebook status updates that relate to real world observations can be fed into the same IoT system. An example is in the EU FP7 project (CityPulse 2013), and this is also further described in Chapter 15, which is on Participatory Sensing (PS).

Looking towards the applications and services in the IoT, we see that the application opportunities are open-ended, and only imagination will set the limit of what is achievable. Starting from typical M2M applications, one can see application domains emerging that are driven from very diverse needs from across industry, society, and people, and can be of both local interest and global interest. Applications can focus on safety, convenience, or cost reduction, optimizing business processes, or fulfilling various requirements on sustainability and assisted living. Listing all possible application segments is futile, as is providing a ranking of the most important ones. We can point to examples of emerging application domains that are driven by different trends and interests (Figure 2.4). As can be seen, they are very diverse and can include applications like urban agriculture, robots and food safety tracing, and we will give brief explanations of what these three examples might look like.

Urban Agriculture. Already today, more than 50% of the world's population lives in urban areas and cities. The increased attention on sustainable living includes reducing transportation, and in the case of food production, reducing the needs for pesticides. The prospect of producing food at the place where it is consumed (i.e. in urban areas) is a promising

Consumer electronics	Automotive Transport	Retail Banking	Environmental	Infrastructures
• Connected gadgets • Wearables • Robotics • Participatory sensing • Social Web of Things	• Autonomous vehicles • Multimodal transport	• Micro payments • Retail logistics • Product life-cycle info • Shopping assistance	• Pollution • Air, water, soil • Weather, climate • Noise	• Buildings and Homes • Roads, rail

Utilities	Health Well-being	Smart Cities	Process industries	Agriculture
• Smart Grid • Water management • Gas, oil and renewables • Waste management • Heating, Cooling	• Remote monitoring • Assisted living • Behavioral change • Treatment compliance • Sports and fitness	• Integrated environments • Optimized operations • Convenience • Socioeconomics • Sustainability • Inclusive living	• Robotics • Manufacturing • Natural resources • Remote operations • Automation • Heavy machinery	• Forestry • Crops and farming • Urban agriculture • Livestock and fisheries

FIGURE 2.4

Emerging IoT applications.

example. By using IoT technologies, urban agriculture could be highly optimized. Sensors and actuators can monitor and control the plant environment and tailor the conditions according to the needs of the specific specimen. Water supply through a combination of rain collection and remote feeds can be combined on demand. City or urban districts can have separate infrastructures for the provisioning of different fertilizers. Drainage can be provided so as not to spoil crops growing on facades and rooftops of buildings, as well as to take care of any recyclable nutrients. Weather and light can be monitored, and necessary blinds that can shield and protect, as well as create greenhouse microclimates, can be automatically controlled. Fresh air generated by plants can be collected and fed into buildings, and tanks of algae that consume waste can generate fertilizers. A vision of urban agriculture is to be a self-sustaining system. Urban agriculture can be a mix of highly industrialized deployments with vertical greenhouses (Plantagon 2013), and collective efforts by individuals in apartments by the use of more do-it-yourself style equipment (Bitponics 2013).

Robots. The mining industry is undergoing a change for the future. Production rates must be increased, cost per produced unit decreased, and the lifetime of mines and sites must be prolonged. In addition, human workforce safety must be higher, with fewer or no accidents, and environmental impact must be decreased by reducing energy consumption and carbon emissions. The mining industry answer to this is to turn each mine into a fully automated and controlled operation. The process chain of the mine involving blasting, crushing, grinding, and ore processing will be highly automated and interconnected. The heavy machinery used will be remotely controlled and monitored, mine sites will be connected, and shafts monitored in terms of air and gases. As up to 50% of energy consumption in a mine can come from ventilation, energy savings can be done by very precise ventilation where the diesel vehicles are operating, and sensors in the mine can provide information about the location of the machines. The trend is also that local control rooms will be replaced by larger control rooms at the corporate headquarters. Sensors and actuators to remotely control both the sites and the massive robots in terms of mining machines for drilling, haulage, and processing are the instruments to make this happen. Companies like Rio Tinto (2012) with their Mine of the Future program, as well as ABB (2013), drive this development.

Food Safety. After several outbreaks of food-related illnesses in the U.S., the U.S. Food and Drug Administration (USFDA) created its Food Safety and Modernization Act (FSMA 2011). The main objective with

FSMA is to ensure that the U.S. food supply is safe. Similar food safety objectives have also been declared by the European Union and the Chinese authorities. These objectives will have an impact across the entire food supply chain, from the farm to the table, and require a number of actors to integrate various parts of their businesses. From the monitoring of farming conditions for plant and animal health, registration of the use of pesticides and animal food, the logistics chain to monitor environmental conditions as produce is being transported, and retailers handling of food — all will be connected. Sensors will provide the necessary monitoring capabilities, and tags like radio frequency identification (RFID) will be used to identify the items so they can be tracked and traced throughout the supply chain. The origin of food can also be completely transparent to the consumers.

As can be seen by these very few examples, IoT can target very point and closed domain-oriented applications, as well as very open and innovation driven applications. Applications can stretch across an entire value chain and provide lifecycle perspectives. Applications can be for business-to-business (B2B) as well as for business-to-consumer (B2C), and can be complex and involve numerous actors, as well as large sets of heterogeneous data sources.

We will progress to see how IoT is driven by a set of diverse needs, and how based on those needs, one can arrive at a set of different needed, recurring capabilities. We will also see how different technologies emerge that will enable building IoT, as well as a generalized model, or architecture, for how to build different target IoT solutions.

2.3 M2M towards IoT — the global context

M2M solutions have been around for decades and are quite common in many different scenarios. While the need to remotely monitor and control assets — personal, enterprise or other — is not new, a number of concurrent things are now converging to create drivers for change not just within the technology industry, but within the wider global economy and society. Our planet is facing massive challenges — environmental, social, and economic. The changes that humanity needs to deal with in the coming decades are unprecedented, not because similar things have not happened before during our common history on this planet, but because many of them are happening at the same time. From constraints on natural resources to a reconfiguration of the world's economy, many people are looking to technology to assist with these issues.

Essentially, therefore, a set of *megatrends* are combining to create *needs and capabilities*, which in turn produce a set of *IoT Technology and Business Drivers*. This is illustrated in Figure 2.5.

A megatrend is a pattern or trend that will have a fundamental and global impact on society at a macro level over several generations. It is something that will have a significant impact on the world in the foreseeable future. We here imply both game changers as challenges, as well as technology and science to meet these challenges. A full description of megatrends is beyond the scope of this book, and interested readers are directed to the many excellent books and reports available on this topic, including publications from the National Intelligence Council (NIC 2012), European Internet Foundation (EIF 2009), Frost & Sullivan (Singh 2012), and McKinsey (McKinsey 2013). In the following section, we focus on the megatrends that have implications for IoT. For the sake of simplicity, we also provide Table 2.1 as a summary of the main game changers, technology and science trends, capabilities, and implications for IoT.

2.3.1 Game changers

The game changers come from a set of social, economic, and environmental shifts that create pressure for solutions to address issues and problems, but also opportunities to reformulate the manner in which our world faces them. There is an extremely strong emerging demand for monitoring, controlling, and understanding the physical world, and the game changers are working in conjunction with technological and scientific advances. The transition from M2M towards IoT is one of the key facets of the technology evolution required to face these challenges. We outline some of these

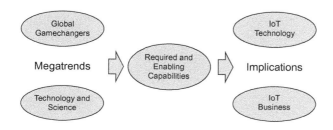

FIGURE 2.5

Megatrends, capabilities, and implications.

Table 2.1 A Summary of Megatrends, Capabilities, and IoT Implications

Megatrends

Global gamechangers
- Natural resource constraints
- Economic shifts
- Changing demographics
- Socioeconomic expectations
- Climate change
- Environmental impacts
- Safety and security
- Urbanization

Technology and Science
- Information and Communication Technologies
- Material science
- Complex and advanced machinery
- Energy production and storage

Capabilities

Required capabilities
- Integrated infrastructures
- Asset-to-expert system integration
- Large scale monitor and control
- Autonomous operations
- Complex remote control
- Workforce offloading
- Domain expertise inside systems
- Visualization
- Data and service exposure
- Advanced analytics
- Increasing levels of security
- Cross value chain integration
- Cost rationalization

Enabling technologies
- Sensing and actuation
- Embedded computing
- Ubiquitous connectivity
- Data processing and storage
- Intelligent software
- Virtualization and cloud
- Application development

Implications

Technology
- Vertical to horizontal systems
- Application independent devices
- Technology consolidation
- IP and Web enabled
- Open software development
- Exposure APIs
- Software enabling architectures
- Cloud deployments
- Intelligence and automation

Business
- Open and innovation driven
- Cloud and as-a-Service delivered
- B2B2C
- Service oriented
- Developer community reach
- Long tail empowerment
- Marketplaces of data and services
- New market roles/value systems
- Cross domain integration
- Commoditized devices
- Application and user driven

more globally significant game changers below, and their relationship to IoT:

- **Natural Resource Constraints**. The world needs to increasingly do more with less, from raw materials to energy, water or food, the growing global population and associated economic growth demands put increasing constraints on the use of resources. The use of IoT to increase yields, improve productivity, and decrease loss across global supply chains is therefore escalating.
- **Economic Shifts**. The overall economy is in a state of flux as it moves from the post-industrial era to a digital economy. One example of this is found in the move from product-oriented to service-oriented economies. This implies a lifetime responsibility of the product used in the service offering, and will in many cases require the products to be connected and contain embedded technologies for gathering data and information. At the same time, there are fluctuations in global economic leadership. Economies across the globe must handle the evolving nature of these forces. As technology becomes increasingly embedded and more tasks automated, countries need to manage this shift and ensure that M2M and IoT also create new jobs and industries.
- **Changing Demographics**. With increased prosperity, there will be a shift in the demographic structures around the world. Many countries will need to deal with an aging population without increasing economic expenditure. As a result, IoT will need to be used, for example, to help provide assisted living and reduce costs in healthcare and emerging "wellcare" systems.
- **Socioeconomic Expectations**. The global emerging middle class results in increasing expectations on well-being and Corporate Social Responsibility. Lifestyle and convenience will be increasingly enabled by technology as the same disruption and efficiency practices evident in industries will be applied within people's lives and homes as well.
- **Climate Change and Environmental Impacts**. The impact of human activities on the environment and climate has been long debated, but is now in essence scientifically proven. Technology, including IoT, will need to be applied to aggressively reduce the impact of human activity on the earth's systems.
- **Safety and Security**. Public safety and national security becomes more urgent as society becomes more advanced, but also more vulnerable. This has to do both with reducing fatalities and health as well as crime prevention, and different technologies can address a number of the issues at hand.

- **Urbanization**. We see the dramatic increase in urban populations and discussions about megacities. Urbanization creates an entirely new level of demands on city infrastructures in order to support increasing urban populations. IoT technologies will play a central role in the optimization for citizens and enterprises within the urban realm, as well as providing increased support for decision-makers in cities.

2.3.2 General technology and scientific trends

Technological and scientific advances and breakthroughs are occurring across a number of disciplines at an increasing pace. Below is a brief description of the science and technology advances that have a direct relevance to IoT. The trends in the ICT sector are described separately in the subsequent section, as it is central for this book.

Material Science has a large impact across a vast range of industries, from pharmaceutical and cosmetics to electronics. MicroElectroMechanical Systems (MEMS) can be used to build advanced micro-sized sensors like accelerometers and gyroscopes. Emerging flexible and printable electronics will enable a new range of innovations for embedding technology in the real world. New materials provide different methods to develop and manufacture a large range of different sensors and actuators, as well being used in applications for environmental control, water purification, etc. Additionally, we will see other innovative uses such as smart textiles that will provide the capability to produce the next generation of wearable technologies. From an IoT perspective, these advances in material science will see an increasing range of applications and also a broader definition of what is meant by a sensor.

Complex and Advanced Machinery refers to tools that are autonomous or semi-autonomous. Today they are used in a number of different industries; for example, robots and very advanced machinery is used in different harsh environments, such as deep-sea exploration, or in the mining industry in solutions such as Rio Tinto's Mine of the Future™ (Rio Tinto 2012). Advanced machines have many modalities, and operate with a combination of local autonomous capabilities as well as remote control. Sensing and actuation are key technologies, and local monitor-control loops for routine tasks are required in addition to reliable communications for remote operations. Often such solutions require real-time characteristics. These systems will continue to evolve and automate tasks today performed by humans — even self-driving cars have started to make headlines thanks to Google.

Energy Production and Storage is relevant to IoT for two reasons. Firstly, it relates to the global interest of securing the availability of electricity while reducing climate and environmental impacts. Smart Grids, for example, imply micro-generation of electricity using affordable photovoltaic panels. In addition, smart grids also require new types of energy storage, both for the grid itself and for emerging technologies such as Electric Vehicles (EVs) that rely on increasingly efficient battery technologies. Secondly, powering embedded devices in Wireless Sensor Networks (WSNs) will increasingly rely on different energy harvesting technologies and also rely on new miniaturized battery technologies and ultra capacitors. As these technologies improve, IoT will be applicable in a broad range of scenarios that need long battery life.

2.3.3 Trends in information and communications technologies

While significant advances in the fields of Material Science, Advanced and Complex Machinery, and Energy Production and Storage will have an impact on IoT, first and foremost, ICT advances will drive the manner in which these solutions are provided as they are the core enabling factors behind M2M and IoT. Ever since the development of integrated circuits during the late 1950s and early 1960s, these technologies have had an increasing impact on enterprises and society. The increasing rates of change have led to a situation where it is now cheap enough to "sensor the planet."

Today, **sensors, actuators, and tags function as the digital interfaces** to the physical world. Small-scale and cheap sensors and actuators provide the bridge between the physical realm and ICT systems. Tags using technologies such as RFID provide the means to put electronic identities on any object, and can be cheaply produced.

Embedded processing is evolving, not only towards higher capabilities and processing speeds, but also extending towards the smallest of applications. There is a growing market for small-scale embedded processing such as 8-, 16-, and 32-bit microcontrollers with on-chip RAM and flash memory, I/O capabilities, and networking interfaces such as IEEE 802.15.4 that are integrated on tiny System-on-a-Chip (SoC) solutions. These enable very constrained devices with a small footprint of a few mm^2 and very low power consumption (in the milli- to micro-Watt range), yet are still capable of hosting an entire TCP/IP stack including a small web server.

Instant access to the Internet is available virtually everywhere today, mainly thanks to wireless and cellular technologies and the rapid deployment of cellular 3G and 4G or Long Term Evolution (LTE) systems on a global scale. These systems provide ubiquitous and relatively cheap connectivity with the right characteristics for many applications, including low latency and the capacity to handle large amounts of data with high reliability. Existing technologies can be further complemented with last-hop technologies such as IEEE 802.15.4, Bluetooth Low Energy, and Power Line Communication (PLC) solutions to reach even the most cost-sensitive deployments and tiniest devices. Technologies like 6LoWPAN allow IP connectivity to be provided end-to-end, stretching into the capillary network domain, and legacy and proprietary protocols like ZigBee PRO can be avoided with the benefit of IP and the web anywhere. 3GPP are also extending LTE towards the lower end of the scale, providing very low power extensions targeting specific IoT applications, also for more constrained devices. Network access technologies are further described in Section 5.2.

Software architectures have undergone several evolutions over the past decades, in particular with the increasing dominance of the web paradigm. A description of the evolution of software is beyond the scope of this book; here we instead look at those angles of software that are critical for successful IoT solutions. From a simplistic perspective, we can view software development techniques from what were originally closed environments towards platforms, where Open APIs provide a simple mechanism for developers to access the functionality of the platform in question (e.g. Microsoft Windows). Over time, these platforms, due to the increasing use and power of the Internet, have become open platforms — ones that do not depend on certain programming languages or lock-in between platform developers and platform owners.

Software development has started applying the **web paradigm and using a service-oriented approach (SOA)**. By extending the web paradigm to IoT devices, they can become a natural component of building any application and facilitate an easy integration of IoT device services into any enterprise system that is based on the SOA (e.g. that uses web services or RESTful interfaces). IoT applications can then become technology and programming language independent. This will help boost the IoT application development market. A key component in establishing the application development market is Open APIs.

Open APIs, in the same way that they have been critical to the development of the web, will be just as important to the creation of a successful IoT market, and we can already see developments in this space. Put simply, Open APIs relate to a common need to create a market *between* many companies, as is the case in the IoT market. Open APIs permit the creation of a fluid industrial platform, allowing components to be combined together in multiple different ways by multiple developers with little to no interaction with those who developed the platform, or installed the devices. As we will discuss in more detail in Chapter 3, it is impossible for one company to guess what will be successful or liked by all of the customer segments associated with IoT. Open APIs are the market's response to this uncertainty; choice of how to combine components is left to developers who are able to merely pick up the technical description and combine them together.

Without Open APIs, a developer would need to create contracts with several different companies in order to get access to the correct data to develop the application. The transaction costs associated with establishing such a service would be prohibitively expensive for most small development companies; they would need to establish contracts with each company for the data required, and spend time and money on legal fees and business development with each individual company. Open APIs remove the need to create such contracts, allowing companies to establish "contracts" for sharing small amounts of data with one another and with developers dynamically, without legal teams, without negotiating contracts, and without even meeting one another. Open APIs therefore reduce the transaction costs associated with establishing a new market boundary (Mulligan 2011), mitigating the risk of development and help to establish a market for innovative capacity, which encourages creativity and application development.

Meanwhile, within ICT, **virtualization** has many different facets and has gained a lot of attention in the past few years, even though it has been around for a rather long time. The **cloud computing** paradigm, with different *as a Service* models, is one of the greatest aspects of the evolution of ICT for IoT as it allows virtualized and independent execution environments for multiple applications to reside in isolation on the same hardware platform, and usually in large data centers.

Cloud computing allows elasticity in deployment of services and enables reaching long-tail applications in a viable fashion. It can be used to avoid in-house installations of server farms and associated dedicated IT service operations staff inside companies, thus enabling them to focus on their core business. Cloud computing also has the benefit of easing different businesses to interconnect if they are executing on the same platform. Handling of, for example, Service Level Agreements (SLAs) is easily facilitated with a high degree of control in a common virtualized environment. Cloud computing is also a key enabler when moving from a product-oriented offering to a service-oriented offering due to elasticity permitting companies to "pay-as-you-grow."

Closely related to the topic of data centers, **data processing and intelligent software** will have an increasing role to play in IoT solutions. A popular concept now is **big data**, which refers to the increasing number and size of data sets that are available for companies and individuals to collect and perform analysis on. Built on large-scale computing, data storage, in-memory processing, and analytics, big data is intended to find insights in the massive data sets produced. Naturally, these technologies are therefore key enablers for IoT, as they allow the collation and aggregation of the massive datasets that devices and sensors are likely to produce.

IoT is unique in comparison to other big data applications such as social media analysis, however, in that even the smallest piece of data can be critical. Take, for example, a sensor solution implemented to ensure that a large-scale engineering project does not cause subsidence in a residential area while drilling a tunnel beneath the ground. The data collected from a vast quantity of sensors will help with the overall management of the project and ensure the health and safety of those working on it during the several months that it is ongoing. It might only be a tiny piece of data from one sensor, however, that indicates a shift due to tunneling that may mean the collapse of a building on the surface. Whereas the aggregation of the data from all sensors can be usefully analyzed at intervals, the data related to subsidence and possible collapse of a building is critical and required in real-time.

Therefore, IoT data typically also involves numerous and very different and heterogeneous sources, but also numerous and very different usages of the data. The analysis of IoT data may therefore be viewed as a complex set of interactions related to *time* (i.e. when the data is received) and *relevance* (i.e. the overall relevance of the piece of data to the question in hand). Managing these interactions is critical to the success of IoT solutions.

Decision support or even decision-making systems will therefore become very important in different application domains for IoT, as will the set of tools required to process data, aggregate information, and create knowledge. Knowledge representation across domains and heterogeneous systems are also important, as are semantics and linked-data. As a result, we can expect to see an increased usage of cognitive technologies and self-learning systems.

A fundamental addition to the data aspect of IoT is the dimension represented by **actionable services as realized by actuators**. There is a duality in sensing and actuation in terms of fusion and aggregation. Where data analytics is employed to find insights basically by aggregation, one can consider complex multimodal actuation services that need to be resolved down to the level of individual atomic actuation tasks. IoT also calls for intelligence in the form of closed control loops of sensing and actuation, which can be simple or very complex. This duality and the closed control aspect will put new requirements on technologies that stretch the boundaries from what can be achieved based on data-only oriented technologies used in the prevalent approach to "big data."

As is illustrated in this chapter, the IoT market holds incredible promise for solving big problems for industry, society, and even individuals. One key thing to note, however, is the tremendous complexity that such systems need to handle in order to function efficiently and effectively. Partnerships and alliances are therefore critical − no one company will be able to produce all the technology and software required to deliver IoT solutions. Moreover, no one company will be able to house the innovative capacity required to develop new solutions for this market. IoT solutions bring together devices, networks, applications, software platforms, cloud computing platforms, business processing systems, knowledge management, visualization, and advanced data analysis techniques. This is quite simply not possible at scale without significant levels of **system integration and standards development**.

This section discussed the global megatrends associated with technology and society. The following section contains a discussion of the capabilities that are required and delivered due to the IoT.

2.3.3.1 Capabilities
As illustrated in previous sections, there are several recurring characteristics of ICT required to develop IoT solutions. These capabilities address several aspects such as cost efficiency, effectiveness and convenience; being lean and reducing environmental impact; encouraging innovation; and in general applying technology to create more intelligent systems, enterprises, and societies. The aforementioned ICT developments provide us with a rich toolbox to address these different aspects in general, and as part of that, IoT in particular. In the following sections we outline how these required capabilities, driven by global megatrends, can be met through the use of the enabling technologies.

While M2M today targets specific problems with tailored, siloed solutions, it is clear that emerging IoT applications will address the much more complex scenarios of large-scale distributed monitor and control applications. IoT systems are multimodal in terms of sensing and control, complex in management, and distributed across large geographical areas. For example, the new requirements on Smart Grids involve end-to-end management of energy production, distribution, and consumption, taking into consideration needs from Demand Response, micro-generation, energy storage, and load balancing. Industrialized agriculture involving automated irrigation, fertilization, and climate control is another example. We see clearly here heterogeneity across sensor data types, actuation services, underlying communication systems, and the need to apply intelligent software to reach various Key Performance Indicators (KPI).

Take, for example, Smart City solutions: here there is a clear need for integration of multiple disparate infrastructures such as utilities, including district heating and cooling, water, waste, and energy, as well as transportation such as road and rail. Each of these infrastructures has multiple stakeholders and separate ownership even though they operate in the same physical spaces of buildings, road networks, and so on. The optimization of entire cities requires the opening up of data and information, business processes, and services at different levels of the disjoint silos, creating a common fabric of services and data relating to the different infrastructures. This integration of multiple infrastructures will drive the need for a horizontal approach at the various levels of the system, for instance, at the resource level where data and information is captured by devices, via the information level, up to the knowledge and decision level. We cover these issues in more detail in Chapter 14: Smart Cities.

Meanwhile, advanced remotely operated machinery, such as drilling equipment in mines or deep sea exploration vessels, will require real-time control of complex operations, including various degrees of autonomous control systems. This places new requirements on the execution of distributed application software and real-time characteristics on both the network itself, as well as a need for flexibility in where application logic is executed.

IoT will allow more assets of enterprises and organizations to be connected, thus allowing a tighter and more prompt integration of the assets into business processes and expert systems. Simple machines can be used in a more controlled and intelligent manner, often called "Smart Objects." These connected assets will generate more data and information, and will expose more service capabilities to ICT systems. Managing the complexity of information and services becomes a growing barrier for the workforce, and places a high focus on using analytics tools of various kinds to gain insights. These insights, combined with domain-specific knowledge, can help the decision process of humans as individuals or professionals via decision support systems and visualization software.

As society operations involve a large number of actors taking on different roles in providing services, and as enterprises and industries increasingly rely on efficient operations across ecosystems, cross value chain and value system integration is a growing need. This requires technologies and business mechanisms that enable operations and information sharing across supply chains. Even industry segments that have been entirely unconnected will connect due to new needs; an example is the introduction of EVs. EVs are enabled by the new battery and energy storage technologies, but also require three separate elements to be connected — cars, road infrastructure via charging poles, and the electricity grid. In addition, there are new charging requirements that are created by the use of EVs that need new means for billing, and in turn placing new requirements on the electricity grid itself.

These sorts of collaboration scenarios will become increasingly important as industries, individuals, and government organizations work together to solve complex problems involving multiple stakeholders. This places an emphasis on the openness and exposure of services and information at different levels. What is important is to be able to share information and services across organizations in the horizontal dimension, as well as being able to aggregate and combine services and information to reach higher degrees of refinement and values in the vertical dimension. The open and collaborative nature of IoT means methods are required to publish and discover data and services, as well as means to achieve semantic interoperability, but also that care needs to be given to trust, security, and privacy. It also dramatically increases the required capability of system integration and the management of large-scale complex systems across multiple stakeholders and multiple organizational boundaries.

As we come to increasingly rely on ICT solutions to monitor and control assets, physical properties of the real world require not just increased levels of cybersecurity, but what can be referred to as cyber-physical security. In the use of the Internet today, it is possible to exact financial damage via breaking into information technology (IT) systems of companies or bank accounts of individuals. Individuals, meanwhile, can face social damages from people hacking social media accounts. In an IoT, where it is possible to control assets (e.g. vehicles or moveable bridges), severe damage to property, or even loss of life, is possible. This raises requirements for trust and security to be correctly implemented in IoT systems.

2.3.4 Implications for IoT

Having gained a better understanding of capabilities needed, as well as how technology evolution can support these needs, we can note plausible implications on both the technology and business perspectives.

There is already a trend of moving away from vertically oriented systems, or application-specific silos, towards a horizontal systems approach. We see that in the standardization work of the ETSI Technical Committee M2M (ETSI M2M 2013), and now also in the oneM2M project partnership organization, both covered in more detail in Part II of this book. The work in these organizations is primarily based on identifying a common set of service capabilities that are application independent, but they also identify reference points to underlying communication network services as well as reference points to M2M devices.

The use of the TCP/IP stack towards IoT devices represents another horizontal point in an M2M and IoT system solution, and is something driven by organizations like the IETF and the IP for Smart Objects (IPSO) Alliance. In the M2M device area, there is an emerging consolidation of technologies where solutions across different industry segments traditionally rely on legacy and proprietary technologies. Currently within industry segments there is technology fragmentation, one example being Building and Home Automation and Control with legacy technologies like BACnet, Lonworks, KNX, Z-Wave, and ZigBee. An example of this consolidation entering the legacy domain is ZigBee IP (ZigBee Alliance 2013c), where TCP/IP is used in the Smart Energy Profile (SEP) 2.0 (ZigBee Alliance 2013b).

In such situations, where there is a requirement for integration across multiple infrastructures and of a large set of different devices, as well as data and information sharing across multiple domains, there is a clear benefit from a horizontal systems approach with at least a common conceptual interoperability made available, and a reduced set of technologies and protocols being used.

As mentioned previously, M2M is point problem-oriented, resulting in point solutions where devices and applications are highly dedicated to solving a single task. M2M devices are for this reason many times highly application-specific, and reuse of devices beyond the M2M application at hand is difficult, if at all possible. With the increasing requirements to gather information and services from various sources, and to be able to have greater flexibility and variety in IoT applications, devices can no longer be application-specific in the same manner as for M2M. Benefits will be achieved if an existing device can be used in a variety of applications, and likewise if a specific application can use a number of different deployed devices. Here we see a shift from application-specific devices towards application-independent devices. As also mentioned, clear benefits come from relying on the web services paradigm, as it allows easy integration in SOAs and attracts a larger application developer community.

Even though M2M has been around for many years, recent years have seen a tremendous interest in M2M across industries, primarily the telecom industry. This comes from the fact that both devices and connectivity have become viable for many different applications, and M2M today is centered on devices and connectivity. For IoT there will be a shift of focus away from device- and connectivity-centricity towards services, data, and intelligence.

2.3.5 Barriers and concerns

With the explained transformations in moving from M2M towards IoT, which involves many opportunities, we should not forget that some new concerns and barriers will also be raised.

With the IoT, the first concern that likely comes to mind is the compromise of privacy and the protection of personal integrity. The use of RFID tags for tracing people is a raised concern. With a massive deployment of sensors in various environments, including in smartphones, explicit data and information about people can be collected, and using analytics tools, users could potentially be profiled and identified even from anonymized data.

The reliability and accuracy of data and information when relying on a large number of data sources that can come from different providers that are beyond one's own control is another concern. Concepts like Provenance of Data and Quality of Information (QoI) become important, especially considering aggregation of data and analytics. As there is a risk of relying on inaccurate or even faulty information in a decision process, the issue of accountability, and even liability, becomes an interest. This will require new technology tools; an example effort includes the work on QoI related to both sensor data and actuation services in the EU FP7 project SENSEI (SENSEI 2013).

As has already been mentioned, the topic of security has one added dimension or level of concern. Not only are today's economical or social damages possible on the Internet, but with real assets connected and controllable over the Internet, damage of property as well as people's safety and even lives become an issue, and one can talk about cyber-physical security.

Not a concern, but a perceived barrier for large-scale adoption of IoT is in costs for massive deployment of IoT devices and embedded technologies. This is not only a matter of Capital Expenditure (CAPEX), but likely more importantly a matter of Operational Expenditure (OPEX). From a technical perspective, what is desired is a high degree of automated provisioning towards zero-configuration. Not only does this involve configuration of system parameters and data, but also contextual information such as location (e.g. based on Geographic Information System (GIS) coordinates or room/building information).

These different concerns and barriers have consequences not only on finding technical solutions, but are more importantly having consequences also on business and socioeconomic aspects as well as on legislation and regulation. The market perspective is further covered in Chapter 3.

2.4 A use case example

In order to understand how a specific problem can be addressed with M2M and IoT, respectively, we provide a fictitious illustrative example. Our example takes two different approaches towards the solution, namely an M2M approach and an IoT approach. By that, we want to highlight the potential and benefits of an IoT-oriented approach over M2M, but also indicate some key capabilities that will be required going beyond what can be achieved with M2M. Our example is taken from personal well-being and health care.

Studies from the U.S. Department of Health and Human Services have shown that close to 50% of the health risks of the enterprise workforce are stress related, and that stress was the single highest risk contributor in a group of factors that also included such risks as high cholesterol, overweight issues, and high alcohol consumption. As stress can be a root cause for many direct negative health conditions, there are big potential savings in human quality of life, as well as national costs and productivity losses, if the factors contributing to stress can be identified and the right preventive measures taken. By performing the steps of stressor diagnosis, stress reliever recommendations, logging and measuring the impacts of stress relievers for making a stress assessment, all in an iterative approach, there is an opportunity to significantly reduce the negative effects of stress.

Measuring human stress can be done using sensors. Two common stress measurements are heart rate and galvanic skin response (GSR), and there are products on the market in the form of bracelets that can do such measurements. These sensors can only provide the intensity of the heart rate and GSR, and do not provide an answer to the cause of the intensity. A higher intensity can be the cause of stress, but can also be due to exercise. In order to analyze whether the stress is positive or negative, more information is needed.

The typical M2M solution would be based on getting sensor input from the person by equipping him or her with the appropriate device, in our case the aforementioned bracelet, and using a smartphone as a mobile gateway to send measurements to an application server hosted by a health service provider. In addition to the heart rate and GSR measurements, an accelerometer in the smartphone measures the movement of the person, thus providing the ability to correlate any physical activity to the excitement measurements. The application server hosts the necessary functionality to analyze the collected data, and based on experience and domain knowledge, provides an indication of the stress level. The stress information can then be made available to the person or a caregiver via smartphone application or a web interface on a computer. The M2M system solution and measured data is depicted in Figure 2.6.

As already pointed out, this type of solution that is limited to a few measurement modalities can only provide very limited (if any) information about what actually causes the stress or excitement. Causes of stress in

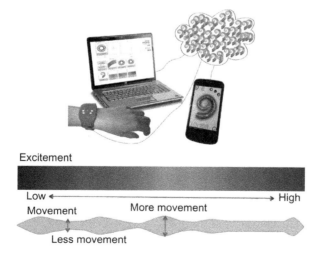

FIGURE 2.6

Stress measurement M2M solution.

(Courtesy of Swedish Institute of Computer Science and Ericsson)

daily life, such as family situation, work situation, and other activities cannot be identified. A combination of the stress measurement log over time, and a caregiver interviewing the person about any specific events at high levels of measured stress, could provide more insights, but this is a costly, labor-intensive, and subjective method. If additional contextual information could be added to the analysis process, a much more accurate stress situation analysis could potentially be performed.

Approaching the same problem situation from an IoT perspective would be to add data that provide much deeper and richer information of the person's contextual situation. The prospect is that the more data is available, the more data can be analyzed and correlated in order to find patterns and dependencies. What is then required is to capture as much data about the daily activities and environment of the person as possible. The data sources of relevance are of many different types, and can be openly available information as well as highly personal information. The resulting IoT solution is shown in Figure 2.7, where we see examples of a wide variety of data sources that have an impact on the personal situation. Depicted is also the importance of having expert domain knowledge that can mine the available information, and that can also provide proposed actions to avoid stressful situations or environments.

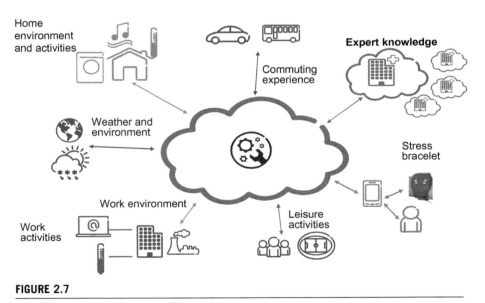

FIGURE 2.7

IoT-oriented stress analysis solution.

(Courtesy of Ericsson)

The environmental aspects include the physical properties of the specific environment, and can be air quality and noise levels of the work environment, or the nighttime temperature of the bedroom, all having impacts on the person's well-being. Work activities can include the amount of e-mails in the inbox or calendar appointments, all potentially having a negative impact on stress. Leisure activities, on the other hand, can have a very positive impact on the level of excitement and stress, and can have a more healing effect than a negative effect. Such different negative and positive factors need to be separated and filtered out; see Figure 2.8 for an example smartphone application that provides stress analysis feedback.

The stress bracelet is in this scenario is just one component out of many. It should also be noted that the actual information sources are very independent of the actual application in mind (i.e. measurement and prevention of negative stress).

By having the appropriate expert knowledge system in place, analytics can be proactive and preventive. By understanding what factors cause negative stress, the system can propose actions to be taken, or even initiate actions automatically. They could be very elementary, such as suggesting to lower the nighttime bedroom temperature a few degrees, but also be

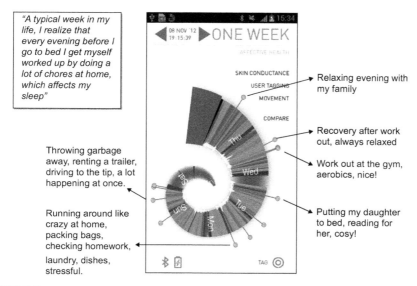

FIGURE 2.8

Stress analysis visualization.

(Courtesy of Swedish Institute of Computer Science and Ericsson)

more complex, such as having to deal with an entire workplace environment.

As this simple example illustrates, an IoT-oriented solution to solving a particular problem could provide much more precision in achieving the desired results. We also observe some of the key features of an IoT solution; in other words, to take many different data sources into account, relying both on sensor-originated data sources, but also other sources that have to do with the physical environment, and then also to rely on both openly available data as well as data that is private and personal. The data sources, such as sensor nodes, should also focus on providing the information and should to the greatest extent be application-independent so that their reuse can be maximized. We also see the central role of analytics and knowledge extraction, as well as taking knowledge into actionable services that can involve controlling the physical environment using actuators. The increased complexity also comes at a cost. The solutions must ensure security and protection of privacy, and the need to deal with data and information of different degrees of accuracy and quality needs to be addressed in order to provide dependable solutions in the end.

2.5 Differing characteristics

To summarize, today's M2M solutions and deployments share a few common characteristics. First of all, any M2M solution is generally focused on solving a problem at a particular point for one company or stakeholder. It does not typically take a broad perspective on solving a larger set of issues or ones that could involve several stakeholders. As a result, most M2M devices are special purpose devices that are application-specific, often down to the device protocols. M2M solutions are therefore also vertical silos with no horizontal integration or connection to adjacent use cases, and are primarily of a B2B-type of operation. M2M applications are built by very specialized developers, and deployed inside enterprises. As M2M has a rather long history, technologies used are very industry-specific, and especially on the device side, technology use is highly fragmented with little or no standards across industries. M2M is also very device- and communication-centric, as both are the two current cornerstones for remote access to assets.

The transition from M2M towards an IoT is mainly characterized by moving away from the mentioned closed-silo deployments towards something that is characterized by openness, multipurpose, and innovation.

Table 2.2 A Comparison of the Main Characteristics of M2M and IoT

Aspect		M2M	IoT
Applications and services		Point problem driven	Innovation driven
		Single application - single device	Multiple applications - multiple devices
		Communication and device centric	Information and service centric
		Asset management driven	Data and information driven
Business		Closed business operations	Open market place
		Business objective driven	Participatory community driven
		B2B	B2B, B2C
		Established value chains	Emerging ecosystems
		Consultancy and Systems Integration enabled	Open Web and as-a-Service enabled
		In-house deployment	Cloud deployment
Technology		Vertical system solution approach	Horizontal enabler approach
		Specialized device solutions	Generic commodity devices
		De facto and proprietary	Standards and open source
		Specific closed data formats and service descriptions	Open APIs and data specifications
		Closed specialized software development	Open software development
		SOA enterprise integration	Open APIs and web development

This transition consists of a few main transformations, namely: moving away from isolated solutions to an open environment; the use of IP and web as a technology toolbox, the current Internet as a foundation for enterprise and government operations; multimodal sensing and actuation; knowledge-creating technologies; and the general move towards a horizontal layering of both technology and business. The main differing characteristics between M2M and IoT highlighted in this chapter are summarized in Table 2.2.

M2M to IoT – A Market Perspective

3.1 Introduction

The increasing interest in M2M and IoT solutions has been driven by the potential large market and growth opportunities. The global environmental sensor and monitoring market, for example, was valued at $11.1 billion in 2010, and is expected to reach $15.3 billion by 2016. The global market for products created with remotely sensed data is predicted to reach $12.4 billion by 2017 (BCC Research 2012), with the global industrial automation market forecast to reach more than $200 billion by 2015 (IMS Research 2013).

The market development for IoT, however, is intricately linked to the technology implemented today, and how this will evolve to provide new economic benefits and value creation opportunities. Chapter 2 outlined these processes from a global technology perspective; this chapter

From Machine-to-Machine to the Internet of Things: Introduction to a New Age of Intelligence.
DOI: http://dx.doi.org/10.1016/B978-0-12-407684-6.00003-6

presents an overview of the market drivers that will see M2M evolve into a full-fledged IoT market. We discuss firstly the concepts behind the drive towards Information-Driven Value Chains (Mulligan 2011), and then provide some real-world examples of how these are emerging today.

3.1.1 Information marketplaces

A key aspect to note between M2M and IoT is that the technology used for these solutions may be very similar — they may even use the same base components — but the manner in which the data is managed will be different. In an M2M solution, data remains within strict boundaries — it is used solely for the purpose that it was originally developed for. With IoT, however, data may be used and reused for many different purposes, perhaps beyond the original intended design, thanks to web-based technologies. As discussed in Chapter 2, M2M solutions will evolve to be able to share greater data with each other and across value chains — or information marketplaces. Data can be shared between companies and value chains in internal information marketplaces. Alternatively, data could be publicly exchanged on a public information marketplace. These marketplaces are based on the exchange of data in order to create information products. This is discussed in more detail in Section 3.1.

While public information marketplaces are generally the vision around IoT, in particular for Smart Cities as discussed in Chapter 14, it is unlikely such marketplaces will become commonplace before trust, risk, security, and insurance for data exchanges are able to be fully managed appropriately.

In the following sections, we therefore focus on the business drivers for delivering M2M solutions and marketplaces that span multiple value chains, rather than publicly traded IoT marketplaces (Figure 3.1).

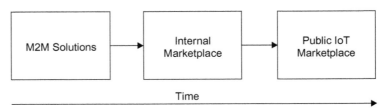

FIGURE 3.1

From M2M to IoT — A Marketplace Perspective.

3.2 **Some definitions**

We do not intend to go into detailed economic theory within this chapter, but we provide some basic working definitions that will provide a working understanding of the market dynamics driving the move from M2M towards IoT, and what business and economic enablers must be put in place in order to drive the overall market.

3.2.1 **Global value chains**

A value chain describes the full range of activities that firms and workers perform to bring a product from its conception to end use and beyond, including design, production, marketing, distribution, and support to the final consumer (Gereffi 2011). A simplified value chain is illustrated in Figure 3.2; it is comprised of five separate activities that work together to create a finalized product.

These activities may be contained within a single firm or divided among different firms (Global Value Chains 2011). Analyzing an industry from a global value chain (GVC) perspective permits understanding of the context of globalization on the activities contained within them by "focusing on the sequences of tangible and intangible value-adding activities, from conception and production to end use. GVC analysis therefore provides a holistic view of global industries – both from the top down and from the bottom up" (Gereffi 2011).

Within the context of the technology industries, GVC analysis is particularly useful as such an analysis can help identify the boundaries between existing industrial structures such as M2M solutions and emerging industrial structures, as seen within the IoT market.

3.2.2 **Ecosystems vs. value chains**

Business Ecosystems, defined by James Moore (Moore 1996), refer to "an economic community supported by a foundation of interacting organizations and individuals ... The economic community produces goods and services of value to customers, who are themselves members of the

FIGURE 3.2

A simplified global value chain.

ecosystem. The member organisms also include suppliers, lead producers, competitors, and other stakeholders. Over time, they co-evolve their capabilities and roles, and tend to align themselves with the directions set by one or more central companies. Those companies holding leadership roles may change over time, but the community values the function of ecosystem leader because it enables members to move toward shared visions to align their investments, and to find mutually supportive roles."

Many people discuss the IoT market as an "ecosystem," with multiple companies establishing loose relationships with one another that then may "piggy back" on larger companies in the ecosystem to deliver products and services to end-users and customers.

While this is a useful description to begin with, a value chain is associated with the creation of value — it is the instantiation of exchange by a certain set of companies *within* an ecosystem. This is an important distinction when we are talking about market creation. A value chain is a useful model to explain how markets create value and how they evolve over time. While a market space composed of only competing value chains will eventually see the overall market value decrease (as they will compete only on price), in an ecosystem, the value chains will complement one another.

In this chapter, we are interested in the creation of a marketplace for IoT data, and we therefore use a GVC analysis; an ecosystem analysis is out of the scope of this book.

3.2.3 Industrial structure

Industrial structure refers to the procedures and associations within a given industrial sector. It is the structure that is purposed towards the achievement of the goals of a particular industry. This is one of the key differences between the M2M and IoT markets — how the industrial structures will be formed around these solutions, despite very similar technology implementations. This is covered in more detail in the following sections.

3.3 M2M value chains

As discussed in Chapter 2, the significant majority of M2M applications have — and will be in the near future — developed for some form of business process optimization. As a result, the majority of organizations will first take an inward-looking approach to business drivers and the reasoning behind why they will implement such solutions. Reasons for using M2M vary from project to project and company to company, but can include things such as cost reductions through streamlined business processes, product quality improvements, and increased health and safety protection for employees. These solutions are generally all internal to a company's business processes and do not included extensive interactions with other parties. Referring back to Figure 3.2, let's take a look at the inputs and outputs of an M2M value chain.

Inputs: Inputs are the base raw ingredients that are turned into a product. Examples could be cocoa beans for the manufacture of chocolate or data from an M2M device that will be turned into a piece of information.

Production/Manufacture: Production/Manufacture refers to the process that the raw inputs are put through to become part of a value chain. For example, cocoa beans may be dried and separated before being transported to overseas markets. Data from an M2M solution, meanwhile, needs to be verified and tagged for provenance.

Processing: Processing refers to the process whereby a product is prepared for sale. For example, cocoa beans may now be made into cocoa powder, ready for use in chocolate bars. For an M2M solution, this refers to the aggregation of multiple data sources to create an information component — something that is ready to be combined with other data sets to make it useful for corporate decision-making.

Packaging: Packaging refers to the process whereby a product can be branded as would be recognizable to end-user consumers. For example, a chocolate bar would now be ready to eat and have a red wrapper with the words "KitKat™" on it. For M2M solutions, the data will have to be combined with other information from internal corporate databases, for example, to see whether the data received requires any action. This data would be recognizable to the end-users that need to use the information, either in the form of visualizations or an Excel spreadsheet.

Distribution/Marketing: This process refers to the channels to market for products. For example, a chocolate bar may be sold at a supermarket, a kiosk, or even online. An M2M solution, however, will have produced an Information Product that can be used to create new knowledge within a corporate environment — examples include more detailed scheduling of maintenance based on real-world information or improved product design due to feedback from the M2M solution.

As mentioned previously, M2M value chains are internal to one company and cover one solution. IoT Value Chains, meanwhile, are about the use and reuse of data across value chains and across solutions.

3.4 IoT value chains

Meanwhile, the move towards IoT — from a value creation perspective — comes with the desire to make some of the data from sensors publicly available as part of an "information marketplace" or other data exchange that allows the data to be used by a broader range of actors rather than just the company that the system was originally designed for. It should be noted that such a marketplace could still be internal to a company or strictly protected between the value chains of several companies. Another alternative is a public marketplace, where data may be treated as a derivative, but such public trading of data is probably a long way from real-world market realization in 2013.

IoT value chains based on data are to some extent enabled by Open APIs and the other open web-based technologies discussed in Chapter 2. Open APIs allow for the knowledge contained within different technical systems to become *unembedded*, creating the possibility for many different economic entities to combine and share their data as long as they have a well-defined interface and description of how the data is formatted. Open APIs in conjunction with the Internet technologies described in Chapter 2 mean that *knowledge is no longer tied to one digital system*. The cognitive and conceptual human skills that were first embedded in semiconductors during the 1950s and 1960s are now decoupled from the specific technological system that was developed to house them. It is this decoupling of technology systems that allows for the creation of information marketplaces. This can initially make the value chain of an IoT solution look significantly more complex than one for a traditional product such as chocolate, but the principles remain the same.

Let's take a closer look at a possible IoT value chain, including an Information Marketplace, illustrated in Figure 3.3.

Inputs: The first thing that is apparent for an IoT value chain is that there are significantly more inputs than for an M2M solution. In Figure 3.3, four are illustrated:

- Devices/Sensors: these are very similar to the M2M solution devices and sensors, and may in fact be built on the same technology. As we will see later, however, the manner in which the data from these devices and sensors is used provides a different and much broader marketplace than M2M does.
- Open Data: Open data is an increasingly important input to Information Value Chains. A broad definition of open data defines it as: "A piece of data is open if anyone is free to use, reuse, and redistribute it — subject

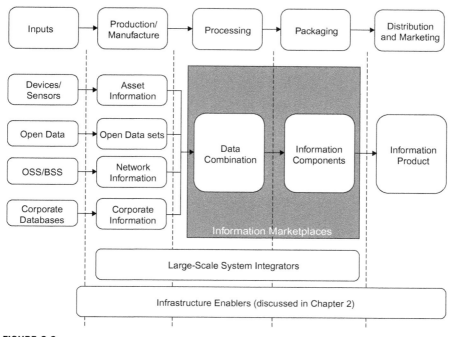

FIGURE 3.3

An Information-Driven Value Chain for IoT.

only, at most, to the requirement to attribute and/or share-alike" (Open Definition 2013). Within the context of this book, we refer to open data as those provided by government and city organizations. Examples include city maps, provided by organizations such as Ordinance Survey in the United Kingdom. Open data requires a license stating that it is open data.

- OSS/BSS: The Operational Support Systems and Business Support Systems of mobile operator networks are also important inputs to information value chains, and are being used increasingly in tightly closed information marketplaces that allow operators to deliver services to enterprises — for example, where phone usage data is already owned by the company in question.
- Corporate Databases: Companies of a certain size generally have multiple corporate databases covering various functions, including supply chain management, payroll, accounting, etc. ... Over the last decades, many of these databases within corporations have been increasingly interconnected using Internet Protocol (IP) technologies.

As the use of devices and sensors increases, these databases will be connected to this data to create new information sources and new knowledge.

Production/Manufacture: In the production and manufacturing processes for data in an IoT solution, the raw inputs described above will undergo initial development into information components and products. Irrespective of input type described above, this process will need to include tagging and linking of relevant data items in order to provide provenance and traceability across the information value chain. Some examples, as illustrated in Figure 3.3, are as follows:

- Asset Information: Asset information may include data such as temperature over time of container during transit or air quality during a particular month. Essentially, this relates to whatever the sensor/device has been developed to monitor.
- Open Data Sets: Open data sets may include maps, rail timetables, or demographics about a certain area in a country or city.
- Network Information: Network information relates to information such as GPS data, services accessed via the mobile network, etc. . . .
- Corporate Information: Corporate information may be, for example, the current state of demand for a particular product in the supply chain at a particular moment in time.

Processing: During the processing stage, data from various sources is mixed together. At this point, the data from the various inputs from the production and manufacture stage are combined together to create information. This process involves the extensive use of data analytics for M2M and IoT solutions and is described in more detail in Chapter 5.

Packaging: After the data from various inputs has been combined together, the packaging section of the information value chain creates information components. These components could be produced as charts or other traditional methods of communicating information to end-users. In addition, however, they could be fed into knowledge management frameworks (discussed in Chapter 5) in order to create not just visualizations of existing information, but to create new knowledge for the enterprise in question.

Both the processing and packaging sections of the Information-Driven Global Value Chain (I-GVC) are where Information Marketplaces will be developed. At this point, data sets with appropriate data tagging and traceability could be exchanged with other economic actors for feeding into their own information product development processes. Alternatively, a company may instead select to exchange information components, which represent a higher level of data abstraction of their corporate information.

Distribution/Marketing: The final stage of the Information Value Chain is the creation of an Information Product. A broad variety of such products may exist, but they fall into two main categories:

- Information products for improving internal decision-making: These information products are the result of either detailed information analysis that allows better decisions to be made during various internal corporate processes, or they enable the creation of previously unavailable knowledge about a company's products, strategy, or internal processes.
- Information products for resale to other economic actors: These information products have high value for other economic actors and can be sold to them. For example, through an IoT solution, a company may have market information about a certain area of town that another entity might pay for (e.g. a real-estate company).

In the following section, we investigate what roles different economic actors will take in the industrial structure for IoT.

3.5 An emerging industrial structure for IoT

Where the technologies of the industrial revolution integrated physical components together much more rapidly, M2M and IoT are about rapidly integrating data and workflows that form the basis of the global economy at increasing speed and precision.

In contrast to fixed broadband technologies, which are limited to implementation in households mainly in the developed world, mobile places consumer electronic goods into the hands of over 4 billion end-users across the globe, and connects billions of new devices into the mobile broadband platform. Concepts such as cloud computing, meanwhile, have the ability to provide low cost access to computational capacity for these billions of end-users via these mobile devices. Combined, these two technologies create a platform that will rapidly redefine the global economy. A new form of value chain is actually emerging as a result — one driven by the creation of information, rather than physical products.

The adoption of the mobile broadband platform is therefore different from previous incarnations of Information and Communication Technology (ICT) industrial platforms as it reshapes not just how economic actors within a value chain interact with one another, but also with employees and the wider economic environment in a similar manner to the technology of the industrial revolution. More importantly perhaps, it changes fundamentally the manner in which *individuals* interact with economic actors in a digital world.

As mentioned in Chapter 2, the need for System Integrators in the communications industries has increased over the decades. With each generation of platform, a new type of system integrator has emerged. For IoT, however, new sets of system integrator capacity are required for two main reasons:

- **Technical:** The factors driving the technical revolution of these industries means that the complexity of the devices in question require massive amounts of R&D; as do semiconductors with large amounts of functionality built into the silicon. Services will require multiple devices, sensors, and actuators from suppliers to be integrated and exposed to developers. Only those companies with sufficient scale to understand the huge number of technologies well enough to integrate them fully on behalf of a customer can handle this technical complexity. While niche integrators will continue to exist, full solutions will be integrated and managed by large companies, or partnerships between vendors.
- **Financial:** Only those companies that are able to capture the added value created in the emerging industrial structure will recoup enough money to re-invest in the R&D required to participate in the systems integration market. It is highly likely that the participants that do not capture part of the integration market will be relegated to "lower" ends of the value chain, producing components as input for other system integrators.

There is in fact a new type of value chain emerging — one where the data gathered from sensors and radio frequency identification (RFID) is combined with information from smartphones that directly identifies a specific individual, their activities, their purchases, and preferred method of communication. This information can be combined in any number of ways to create tailored services of direct relevance to the individual or corporation in question. Search queries can be localized based on where a person is, and advertising can be targeted directly to the end-user in question based on personalized information about their age, level of education, employment, and tastes. While it is perhaps questionable whether the world really needs new methods to advertise goods and services, beneath this development lies a fundamental change in some aspects of the global economy.

Firstly, information about individuals is now captured, stored, processed, and *reused* across many different systems that sit on top of the mobile broadband platform. This data has always existed, but with the increasingly low cost of computing capacity in the form of cloud computing platforms, it is now cheap enough to store this data for an extremely long length of time. It is now possible, therefore, for information about individuals and digital systems to be packaged, bundled, and *exchanged* between economic entities with an ease that has previously been impossible. Value is no longer solely measured through "value-in-use" or "value-in-exchange," but there is now also a "value-in-reuse," specifically because the commodity, data, is not consumed within the processes of production as with previous generations of commodity creation.

Actors that perform this data collection, storage, and processing are forming the basis of what may be viewed as an *Information-Driven Global Value Chain (I-GVC)*, a value chain where the product is information itself.

As an example, a difference in value can be identified in knowing my location when I step off a train in a new city and am looking for a decent cup of coffee. I may choose to activate my smartphone and perform a localized search using my phone's GPS and browser features. Alternatively, I may be happy enough to just walk around until I find a place that I think looks OK. In this case, the value that I as an individual place on my phone knowing my location and assisting me to find a local coffee store is relatively low − personally, I might not value this very highly.

In comparison, however, there is a great deal of value for a coffee company to know that a few hundred women have stepped off a train in search of an espresso. A coffee shop chain would know that it is potentially quite profitable to open a new store there. In addition, understanding the age group of those women, their level of education, and their general tastes would allow the chain to tailor the coffee store to their target market with much greater precision.

Similarly, if I was in a clothing store searching for a new outfit for work, through a combination of information about myself and the RFID tags on the different clothes, I could be guided to the correct clothing selection for my age group, my education level, and also my current employer. Information about what path I take through the store during my search for the clothes could be fed back into an information system that would allow the store to reorganize their floor layouts more effectively, track the clothes that I was interested in, and those which I actually select to try on and purchase. This information can be used to streamline the supply chains of corporations even further than is possible today, and represents the next phase of the impact of communication technologies on the boundaries of the firm within the global economy: companies that share this type of information would be more deeply embedded in one another's workflows, leading to highly concatenated supply chains and a further blurring of the boundaries of the firm within the digital economy. This is illustrated in Figure 3.4.

This streamlining could also be extended into the processes of production, changing orders based on consumer interest in products, and not just their purchasing patterns. This would result in less wasted stock and a much closer understanding of seasonal trends and an increased level of control for those companies working as system integrators. The integration of these data streams allows for concatenation of supply chains not just internally to one company, therefore, but *across industrial boundaries* (Mulligan 2011).

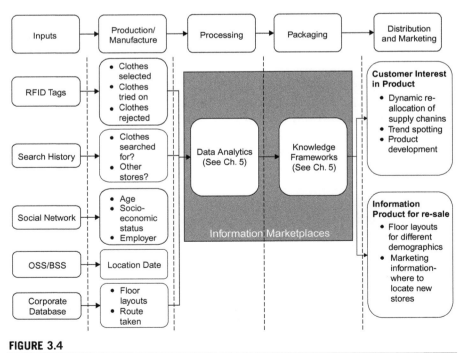

FIGURE 3.4

An Information-Driven Value Chain for Retail.

The level of analysis described above requires *aggregation* of data from many different people and its collation into an *information product*, one that may be used as input into corporate and end-user decision-making processes.

While there is obviously a strong link between the I-GVC and physical goods, it is therefore clear that there is an information product in and of itself, one that relates to the development of aggregation databases that collect data from both sensors and people. Moreover, while the information product is useful to the companies in question, it is not part of their core business to create it. As a result, they look to other actors to develop and create these products for them, which is further driving the creation of a new industrial structure around ICT systems that include IoT and cloud technologies.

The second change in the nature of the economy is the fundamental embedding of human beings into the very foundation of these technology platforms. The most obvious example of this is Google's search engine, which is improved with every search query that is performed using it. Every search that every individual makes is tracked, and every click someone makes through Google's products is recorded and used to refine the algorithms that form the basis of the platform. Without the humans inputting their searches into the Google platform, it quite simply would not exist in the form that it does today.

The broad-scale consumerization of technology, combined with the cheap means of "information production" due to cloud computing, has led to information management systems that are now being developed for end-user consumers, not just enterprises. Social networks such as Facebook and LinkedIn, and content sharing sites such as YouTube or Blogger, allow end-users to store information about their lives in a manner previously not possible. Consumers now store their photos, their contact lists, videos, documents, and financial data online. Ostensibly, this is provided for "free;" end-users receive access to the websites merely through creating an account, logging in, and uploading their data.

Within the capitalist economy, however, no service is ever really free. Companies must pay for the costs of computing resources, even those that are housed in the cloud. While the cost to end-users appears to be zero, they are in fact being charged on a daily basis through the use of their profile data for targeted advertising. In the early days of social networks, for example, this targeted advertising was no worse than the "traditional" direct advertising methods: based on the data provided by the end-user, they would receive an age-appropriate advertisement for their demographic.

With the mobile broadband platform, however, the level of data that can be gathered about end-users is orders of magnitude larger than previously imaginable. My location, level of education, employment status, health records, tax data, credit rating, purchasing patterns, search history, social networks (both private and professional), relationship status, even how often I call my mother are recorded, stored, and *interconnected* in a vast array of disparate systems that are now linked together through the platform of a converged communications industry.

With the addition of new levels of data that it is possible to retrieve about individuals through mobile devices and sensor networks, the emerging ICT platforms are forcing a redefinition of our established understandings of the notion of value both within the communications industries and even beyond these industrial boundaries to all companies and individuals that use these platforms on a daily basis (Mulligan 2011).

The following section outlines the emerging value chain and the roles that must be filled in order to create a flourishing IoT marketplace.

3.5.1 The information-driven global value chain

There are five fundamental roles within the I-GVC that companies and other actors are forming around, illustrated in Figure 3.5:

- Inputs:
  - Sensors, RFID, and other devices.
  - End-Users.
- Data Factories.
- Service Providers/Data Wholesalers.
- Intermediaries.
- Resellers.

3.5.1.1 Inputs to the information-driven global commodity chain

There are two main inputs into the I-GVC:

1. Sensors and other devices (e.g. RFID and NFC).
2. End-users.

Both of these information sources input tiny amounts of data into the I-GVC chain, which are then aggregated, analyzed, repackaged, and exchanged between the different economic actors that form the value chain.

As a result, sensor devices and networks, RFIDs, mobile and consumer devices, Wi-Fi hotspots, and end-users all form part of a network of

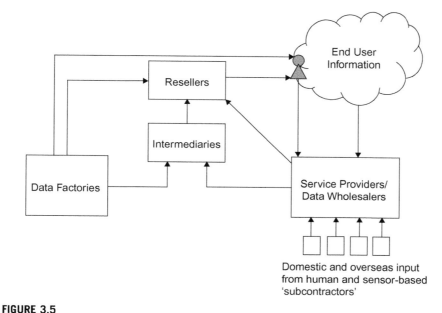

FIGURE 3.5

The Information-Driven Global Value Chain.

"subcontractors" in the value chain, all contributing to the increased value of the information products, which were discussed in Section 3.4.

3.5.1.1.1 Sensors and radio frequency identification

Sensors and RFID are already found in a multitude of different applications worldwide (as discussed in Chapter 2), helping to smooth supply and demand in various supply chains worldwide and gathering climate and other localized data that is then transmitted back to a centralized information processing system. These devices are working as inputs to the I-GVC through the capture and transmission of data necessary for the development of information products.

Smartphones have also been developed that allow mobile devices to interact with sensors and RFID. This allows for a two-way interaction between a mobile terminal and the sensor technology. The data exchanged between the actuator and a mobile terminal may not be readily understood or even useful for the device in question. The data, however, is used as one part of the input to the commodity chain, which uses it to create the information products that are eventually exchanged. In this sense, the sensor networks, and NFC and RFID technologies may be viewed as subcontractors to the I-GVC, workers that constantly gather data for further processing and sale.

3.5.1.1.2 End-users

The second main inputs to the I-GVC are the end-users. Due to the convergence of the computing and mobile broadband platforms, end-users are no longer passive participants in the digital economy, with a role only to purchase those physical products that companies develop and market to them. End-users that choose to use and participate within the digital world are now deeply embedded into the very process of production. Every human that enters a search query into a search engine, every human that agrees to allow the mobile broadband platform to inform a service of their location, every human that uses NFC to allow a bank to establish and confirm their identity are also functioning as subcontractors to the global information systems that form the basis of the I-GVC. In fact, the creation of the I-GVC would not be possible without the contribution of many millions of individuals *worldwide*. This is perhaps the most unique aspect of the I-GVC — there is no national boundary for the contribution of humans to the I-GVC, the data about individuals can be collected from any person in any language, in almost any data format. Each individual's data can be treated as unique within this value chain; in fact, it is the ability to capture the uniqueness of every person that is a key aspect of the I-GVC in comparison to the other commodity chains that are at work within the global economy. Every person worldwide that has to use digital technologies to do their banking, their taxes, their information searches, and to communicate with friends and colleagues, are constantly working on behalf of the I-GVC, contributing their individual profile data and knowledge to the value chain. In the same manner as the actuators constantly gather localized data, humans are now contributing to the development of information products within the I-GVC nearly 24 hours a day.

3.5.1.2 Production processes of the information-driven global value chain

3.5.1.2.1 Data factories

Data factories are those entities that produce data in digital forms for use in other parts of the I-GVC. Many of these companies existed in the predigital era; for example, Ordnance Survey (OS) in the UK has always collected map information from the field, and collated and produced maps for purchase. Previously, such data factories would create paper-based products and sell them to end-users via retailers. With the move to the digital era, however, these companies now also provide this data via digital means; for example, OS now makes maps and associated data available in digital format. Essentially, its business model has not changed significantly — it still produces maps — but its means of delivery of products has changed. Moreover, its products can now be combined, reused, and bundled together with other products by actors in the commodity chain as the foundation of other services. For example, maps from OS can be combined with other data from travel services such as TFL to provide detailed travel applications on mobile devices.

A more complex example is Sveriges Meterologiska och Hydrologiska Institut (SMHI), which provides weather and climate data throughout Sweden. SMHI has a large number of weather stations across Sweden through which it collects weather and environmental information. In addition, it purchases its data from Yr.no, its Norwegian equivalent. SMHI therefore produces raw data, but it also processes the data, and bundles it in different ways based on customer requests and requirements. SMHI functions not only as a data factory, therefore, but also a *reseller*, which is described separately below.

3.5.1.2.2 Service providers/data wholesalers

Service Providers and Data wholesalers are those entities that collect data from various sources worldwide, and through the creation of massive databases, use it to either improve their own information products or sell information products in various forms. Many examples exist; several well-known ones are Twitter, Facebook, Google, etc.... Google "sells" its data assets through the development of extremely accurate, targeted, search-based advertising mechanisms that it is able to sell to companies wishing to reach a particular market. Twitter, meanwhile, through collating streams of "Tweets" from people worldwide, is able to collate customer sentiment about different products and world events, from service at a restaurant to election processes across the globe; through what Twitter refers to as a "data hose," companies and developers can access 50% of end-user Tweets for $360,000 USD per annum.

A new set of data wholesalers is starting to emerge, however: those companies that handle the massive amount of data that is produced by sensor networks and mobile devices worldwide. These companies are collating those transactions that are made by the millions of devices worldwide that utilize communications networks to transmit data. The sheer quantity of data that is being transmitted via actuators and mobile devices will be orders of magnitude higher than previously imagined within the supply chains of multinational companies alone. We cover the technologies associated with these, M2M and Big Data Analytics, in Chapter 5.

3.5.1.2.3 Intermediaries

In the emerging industrial structure of the I-GVC, there is a need for intermediaries that handle several aspects of the production of information products. As mentioned above, there are many privacy and regional issues associated with the collection of personal information. In Europe, the manner in which Facebook collects and uses the data of the individuals that participate in its service may actually be in contravention of European privacy law. The development of databases such as the ones created by Google, Facebook, and Twitter may therefore require the creation of entities that are able to "anonymise" data sufficiently to protect individuals' privacy rights in relevant regional settings. These corporations will provide protection for the consumer that their data is being used in an appropriate manner, i.e. the manner in which the consumer has approved its usage. For example, I may happily share my personalized information about my tastes with a clothing company or music store in order to receive better service, while I may not be happy for my credit rating or tax data to be shared freely with different companies. I would therefore allow an intermediary to act on my behalf, tagging the relevant information in some form to ensure that it was not used in a manner that I had not previously agreed to.

Another reason for an intermediary of this nature is to reduce transaction costs associated with the establishment of a market for many different companies to participate in. As an example, Jasper Wireless acts as an intermediary for the M2M market described in Chapter 3. Through providing a connection point for several different parties in the M2M industry, it acts to expand the uptake of M2M technology.

As discussed in the previous section on service providers/data wholesalers, the amount of data that is being produced is also problematic — even with cloud computing, it is difficult to process the amounts of data produced in the I-GVC. The different types of information products that are to be produced are only of interest to certain types of companies — for example the marketing division of a company may be interested to understand customer sentiment about a particular product within a certain age group. Another company may want to understand what searches are being performed in their local area, while a local authority may wish to use sensor data to obtain real-time data about pollution from local factories. The type of data and style of analysis for each of these information products is fundamentally different, and each requires unique skills — it is highly improbable that one company will be able to handle all of these types of data in one place. It is therefore more likely that different companies, intermediaries, will develop that target their information products to different niche markets. These companies will be able to focus on certain datasets and become specialists within that particular information product field.

The quantity and nature of data being developed into information products also requires a completely new type of intermediary, one that is able to handle the scalability issues and the associated security and privacy questions raised by the use of this data to build products. This is perhaps the most obvious role for operators and network vendors with global services operations to take, as they have many decades of experiences in developing, operating, and maintaining secure systems that scale to millions of users. The average operator network is designed to scale to approximately 100 million end-users. With the advent of data networks for devices (not just human subscribers), operators are now investigating systems that can scale to at least ten-fold that size. Systems that can handle this number of devices and end-users require a huge level of cooperation across industrial structure, and the development of this scale of intermediary will therefore require closer cooperation between equipment manufacturers and service providers at all levels. The I-GVC may therefore also be seen to further blur the boundaries of the firm between the high technology companies that form its basis.

3.5.1.2.4 Resellers

Resellers are those entities that combine inputs from several different intermediaries, combine it together, analyze, and sell it to either end-users or to corporate entities. These resellers are currently rather limited in terms of the data that they are able to easily access via the converged communications platform, but they are indicative of the types of corporate entities that are forming within this space.

One example is BlueKai, which tracks the online shopping behavior of Internet users and mines the data gathered for "purchasing intent" in order to allow advertisers to target buyers more accurately. BlueKai combines data from several sources, including Amazon, Ebay, and Alibaba. Through this data, it is able to identify regional trends, helping companies to identify not just which consumer group to target their goods to, but also which part of the country. As an example, BlueKai is able to identify all those end-users in West Virginia currently searching for a washing machine in a certain price bracket.

3.6 The international-driven global value chain and global information monopolies

Currently within the industrial structure of the converged communications industry, there is a large regional disparity between those companies that produce the infrastructure for the I-GVC and those that make a significant profit from it. Through positioning themselves within the correct part of the GVC, these companies are able to take the lion's share of the profit. Through the breakdown of regional boundaries for collection of data by the development and implementation of a global converged communications infrastructure, these companies are able to enlist every person using a mobile device worldwide as a contributor to the development of their information products — in effect, *every person worldwide is working for these corporations so that they are able to sell aggregated data* for a huge profit. Despite this data being collected from people in every corner of the globe, from the UK, Thailand, Australia, China, and Africa, to even the remotest parts of Kashmir, the surplus value of the mobile broadband platform is currently being captured, developed, and molded into information products, overwhelmingly by U.S. companies. Through being able to collect and analyze data without being restricted by the same level of privacy regulation as in Europe, for example, they are able to create a much better information product. Companies in Europe, Asia, and other parts of the globe are therefore dependent on these companies in order to gain the most appropriate knowledge for their companies' needs. In the same way that the use of IT became a critical success factor for enterprises in the late 1980s and 1990s, the use of information products developed within the I-GVC are becoming critical to securing a competitive advantage in a global market. Companies are therefore compelled to use the most effective information product for their needs.

In effect, the I-GVC, rather than breaking down the digital divide as many have predicted, is in fact leading to a new form of digital discrimination and a new sort of dependency relationship between large multinationals and those participants, or "workers," within the I-GVC. While there may appear to be huge differences between the industrial revolution and the birth of the digital planet in the nature of how workers are treated, in particular with so much being advertised as "free" for end-users, there are in fact many similar parallels in the aggregation of human endeavor in the processes of the accumulation of capital. A multitude of workers contribute to the information products developed, but only a few large corporations capture the surplus value.

This has in fact led to a few interesting discussions within industry about who actually owns this data — Is it the company that provides the service, the service provider that delivers the connectivity, or the end-users themselves? An end-user might potentially be able to receive money from the use of his or her data, a nominal contribution for each time that data is used for creation of an information product. Profit sharing arrangements might even be possible between companies that develop the platform and those that collate the data into product form. The fact remains, however, that it is only those companies with the R&D budgets, the scale, and the global reach necessary to exploit the aggregation of this data that will be able to make significant profits from it.

3.7 Conclusions

This chapter described the drivers for M2M towards IoT from a market perspective. Thanks to open, web-based technologies, IoT solutions will drive the creation of Information Marketplaces that allow the exchange of data between different economic entities within an information value chain. In Chapter 4, we turn to the architectural implementation for IoT solutions.

M2M to IoT – An Architectural Overview

4

CHAPTER OUTLINE

4.1 Building an architecture

The term "architecture" has many interpretations. This chapter provides an introduction to how we use the term "architecture" in this book, and secondly, how it relates to problems, applications of interest, and actual M2M/IoT solutions. A detailed description of theory and philosophy of architectures is beyond the scope of this book; interested readers will find a good example in Rozanski and Woods (2011).

Within this book, architecture refers to the description of the main conceptual elements, the actual elements of a target system, how they relate to each other, and principles for the design of the architecture. A conceptual element refers to an intended function, a piece of data, or a service. An actual element, meanwhile, refers to a technology building block or a protocol. The term "reference architecture" relates to a generalized model that contains the richest set of elements and relations that are of relevance to the domain "Internet of Things."

When looking at solving a particular problem or designing a target application, the reference architecture is to be used as an aid to design an applied architecture, i.e. an instance created out of a subset of the reference architecture. The applied architecture is then the blueprint used to develop the actual system solution (Figure 4.1). The approach taken here is much inspired from the work in the EU FP7 project IoT-A (IoT-A EU FP7 2013).

From Machine-to-Machine to the Internet of Things: Introduction to a New Age of Intelligence.
DOI: http://dx.doi.org/10.1016/B978-0-12-407684-6.00004-8

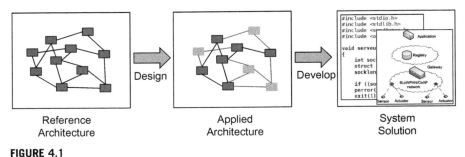

FIGURE 4.1

From a reference architecture to a system solution.

An architecture can be described in several different views to capture specific properties that are of relevance to model, and the views chosen in this book are the functional view, deployment view, process view, and information view. The topic of architecture is the subject of Chapter 6 of this book where more details of the definitions and purpose of an architecture, as well as state of the art examples from M2M and IoT, are provided.

When creating a model for the reference architecture, one needs to establish overall objectives for the architecture as well as design principles that come from understanding some of the desired major features of the resulting system solution. For instance, an overall objective might be to decouple application logic from communication mechanisms, and typical design principles might then be to design for protocol interoperability and to design for encapsulated service descriptions. These objectives and principles have to be derived from a deeper understanding of the actual problem domain, and is typically done by identifying recurring problems or type solutions, and thus by that, extracting common design patterns. The problem domain establishes the foundation for the subsequent solutions. It is common to partition the architecture work and solution work into two domains, each focusing on specific issues of relevance at the different levels of abstraction (Figure 4.2).

The top level of the triangle is referred to here as the "problem domain" ("domain model" in software engineering). The problem domain is about understanding the applications of interest, for example, developed through scenario building and use case analysis in order to derive requirements. In addition, constraints are typically identified as well. These constraints can be technical, like limited power availability in wireless sensor nodes, or non-technical, like constraints coming from legislation or business. These real-world design constraints are covered in more detail in Chapter 9. The lower level is referred to as the solution domain. This is

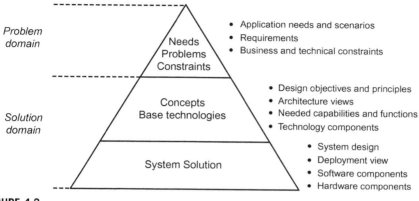

FIGURE 4.2

Problem and Solution domain partitioning.

where design objectives and principles are established, conceptual views are refined, required functions are identified, and where logical partitions of functionality and information are described. Often this is where a logical architecture is defined, or network architecture in the form of a network topology diagram is produced. It is also common to identify suitable technology components such as operating systems and protocols or protocol stacks at this level. The actual system solution is finally captured by a system design that typically results in actual software and hardware components, as well as information on how these are to be configured, deployed, and provisioned.

The next section outlines the design objectives and principles for IoT, the main capabilities, and then an architecture outline. The section also introduces two proposed state-of-the-art examples of architectures from European Telecommunications Standards Institute (ETSI) and IoT-A, respectively, which are covered in chapters 6, 7 and 8.

4.2 Main design principles and needed capabilities

Having introduced some basic concepts in architecture design and how that facilitates the subsequent development of an actual system solution, we turn now to the design principles that underpin architecture design for IoT.

Within existing work for deriving requirements and creating architectures or reference models for IoT and M2M, three primary sources can been identified. Two of them are the larger European 7th Framework Program research projects, SENSEI (2013) and IoT-A (2013), the third being the result of a standardization activity driven by ETSI in their Technical Committee (TC) M2M (ETSI M2M TC 2013). These sources have been selected, as they represent state-of-the-art in terms of creating more complete architectures for the IoT and M2M.

The approach taken in SENSEI was to develop an architecture and technology building blocks that enable a "Real World integration in a future Internet." Key features include the definition of a real world services interface and the integration of numerous Wireless Sensor and Actuator Network deployments into a common services infrastructure on a global scale. The service infrastructure provides a set of services that are common to a vast range of application services and is separated from any underlying communication network for which the only assumption made was that it should be based on Internet Protocol (IP). The architecture relies on the separation of resources providing sensing and actuation from the actual devices, a set of contextual and real world entity-centric services, and the users of the services. SENSEI further relies on an open-ended constellation of providers and users, and also provides a reference model for different business roles. A number of design principles and guidelines are identified, and so is a set of requirements. Finally, the architecture itself contains a set of key functional capabilities.

The telecommunications industry, meanwhile, has focused on defining a common service core for supporting various M2M applications, and that is agnostic to underlying networks in ETSI TC M2M. The approach taken has been to analyze a set of M2M use cases, derive a set of M2M service requirements, and then to specify an architecture as well as a set of supporting system interfaces. Similar to SENSEI, there was a clear approach towards a horizontal system with separation of devices, gateways, communications networks, and the creation of a common service core and a set of applications, all separated by defined reference points.

Finally, the approach taken in IoT-A differs from the two approaches above in the sense that instead of defining a single architecture, a reference architecture is created, captured in what the IoT-A refers to as the Architectural Reference Model (ARM). The vision of IoT-A is, via the ARM, to establish a means to achieve a high degree of interoperability between different IoT solutions at the different system levels of communication, service, and information. IoT-A provides a set of different architectural views, establishes a proposed terminology and a set of Unified Requirements (IoT-A UNI 2013). Furthermore, IoT-A proposes a methodology for how to arrive at a concrete architecture based on use cases and requirements.

Comparing these different approaches, a common feature is the focus on a horizontal system approach. There is a clear separation of the underlying communication networks and related technologies from capabilities that enable services. There is a clear desire to define uniform interfaces towards the devices that provide sensing and actuation, including abstraction of services the devices provide. There is also a clear desire to separate logic that is highly application-specific from logic that is common across a large set of applications. Returning to our previous discussion on understanding trends, needed capabilities, and implications for IoT, we also clearly identified the need for a horizontal approach from a set of different perspectives — the need for horizontal integration across value chains at different levels; the need to integrate multiple infrastructures; the need to reuse existing deployments, to name a few. Taking other key identified features into consideration as well, such as being able to support open-ended service development and providing security and being reliable, we can formulate what can be seen as an overall IoT architecture objective.

The overall design objective of IoT architecture shall be to target a horizontal system of real-world services that are open, service-oriented, secure, and offer trust. Further analyzing both the referenced existing work as well as our conclusions in Chapter 1 on both needed capabilities and direct IoT implications, we can also derive a set of supporting design principles that target different means to fulfill the overall architecture objective. These design principles have a set of interpretations and further expectations on needed technology solutions that we now describe.

Design for reuse of deployed IoT resources across application domains. Deployed IoT resources shall be able to be used in a vast range of different applications. This implies that devices shall be made application-independent and that the basic and atomic services they expose in terms of sensing and actuation shall be done in a (to the greatest extent possible) uniform way. A system design will benefit from providing an abstracted view of these basic underlying services that also are decoupled from the devices that provide the services.

Design for a set of support services that provide open service-oriented capabilities and can be used for application development and execution. What we already have seen in the service layer-oriented M2M standardization in ETSI M2M is the definition of a set of common application-independent service capabilities. These support services shall in general cater to the typical environment of a stakeholder where IoT applications are to be built, such as an open environment, and shall in particular provide support for a few key service capabilities that are central from an IoT perspective. The open environment of IoT will, for instance, require mechanisms for authorized usage of services and resources, authentication, and associated identity management.

The key support services that are required from an IoT perspective include the means to access IoT resources, how to publish and discover resources, tools for modeling contextual information and information related to the real world entities that are of interest, and capabilities that provide different levels of abstracted and complex services. The latter can include data and event filtering and analytics, as well as dynamic service composition and resolution of mixed sensing and actuation. Furthermore, well-defined service interfaces and application programming interfaces (APIs) are required to facilitate application development, as are the appropriate Software Development Kits (SDK).

Design for different abstraction levels that hide underlying complexities and heterogeneities. As we have already seen, typical IoT solutions can involve a large number of different devices and associated sensor modalities, and involve a large set of different actors providing services and information that need to be composed and accessed with different levels of aggregation. A system design will greatly benefit from providing the necessary abstractions both of underlying technologies, data and service representation, as well as granularity of information and services. This will ease the burden of both system integrators and application developers. Again, hiding device-side technologies and providing simple abstractions of the sensing and actuation services is one aspect. Another is the means to perform aggregation of information or knowledge representation. A third, as already discussed, is the requirement to have appropriate knowledge management tools and a means to compose complex services as well as decomposition of complex queries and tasks down to individual and atomic actuation tasks.

Design for sensing and actors taking on different roles of providing and using services across different business domains and value chains. As discussed in Chapter 3 on the IoT Market aspects, as well as in Chapter 2, there are different levels of openness of the business context in which IoT solutions are deployed and running. IoT solutions can be run across a set of departments within an enterprise, or across a set of enterprises in a value system, or even be provided in a truly open environment. The business contexts can then be viewed as no market (entirely intra-organizational), as closed markets (finite and predetermined set of business actors in a specific value system or value chain), or as open markets (undefined and open-ended number of participants). In these different setups there are varying degrees of needed capabilities that address the multi-stakeholder perspective. The first thing that needs to be provided is a set of mechanisms that ensure security and trust. Trust and identity management that refer to the different stakeholders is a fundamental requirement. Authentication and authorization of access to use services as well as to be able to provide services is then a second requirement. The third requirement is the capability to be able to do auditing and to provide accountability so that stakeholders can enforce liability if the need occurs. The next fundamental requirement is to ensure interoperability. This is needed at different levels across the interaction points between the stakeholders. Primary examples are to ensure data and information interoperability on the semantic level, and means to connect business processes across organizational and administrative boundaries. The third fundamental requirement is related to the market perspective, whether the markets are closed or fully open. Mechanisms that provide compensation for used services or data between service users and service providers are needed. As an IoT market can involve everything from trading individual sensor data to aggregated insights and knowledge, compensation and billing mechanisms are needed that can operate on the micro level as well as on more traditional macro levels. An open market environment also calls for means to publish or advertise services, as well as a means for finding services. These different main requirements also provide opportunities for new market roles, such as aggregator roles, broker roles, and clearing houses, all well known in other existing markets.

Design for ensuring trust, security, and privacy. Trust within IoT often implies reliability, which can be both ensuring the availability of services as well as how dependable the services are, and that data is only used for the purposes the end-user has agreed to. One important aspect of dependability is the accuracy of data or information, as you can have multiple sources of IoT data. Concepts like Quality of Information become important, especially considering that a piece of information can very well be accurate enough for one application, but not for another. As has been already mentioned, security and privacy are potential barriers for IoT adoption and represent key areas to address when building solutions. Privacy needs to be ensured by, for example, anonymization of data, seeing that profiling of individuals is not easily done or even made undoable. Still, it is foreseen that authorities and agencies will require support to get access to data and information for the purpose of national security or public safety.

Design for scalability, performance, and effectiveness. IoT deployments will happen on a global scale and are foreseen to involve billions of deployed nodes. Sensor data will be provided with a wide range of different characteristics. Data may be very infrequent (e.g. alarms or detected abnormal events), or may be coming as real-time data streams, all dependent on the type of data needed or based on application needs. Scalability aspects of importance include the large number of devices and amounts of data produced that needs to be processed or stored. Performance includes consideration of mission-critical applications such as Supervisory Control And Data Acquisition (SCADA) systems with extreme requirements on latency, for example.

Design for evolvability, heterogeneity, and simplicity of integration. Technology is constantly changing, and given the nature of IoT deployments where devices and sensor nodes are expected to be operational and in the field for many years, sometimes with lifecycles of over 15 years (e.g. smart meters), IoT solutions must be able to withstand and cater to introduction and use of new technologies as well as handling of legacy deployments. Handling heterogeneity is also important since especially device-oriented technologies used across industries are very different. Means to integrate legacy devices using many different protocols becomes a necessity, and gateways of different types and with different capabilities will be essential to expose capabilities of legacy devices in a uniform manner.

Design for simplicity of management. Again going back to one of the potential barriers for IoT adaptation, simplicity of management is an important capability that needs to be properly taken care of when designing IoT solutions. Autoconfiguration and autoprovisioning are key and well-known means that can ease deployment of IoT devices, and are also very important to lower operating expenditures (OPEX).

Design for different service delivery models. We already know about the clear trends to move from product offerings to a more combined product and service offering in a number of industries, for instance, connected vehicles, and Software as a Service (SaaS) as a delivery model. IoT with the wide span of possible applications clearly benefit from elasticity in deployment of solutions, all to meet the long-tail aspect. Cloud and virtualization technologies play a key enabler role in delivering future IoT services.

Design for lifecycle support. The lifecycle phases are: planning, development, deployment, and execution. Management aspects include deployment efficiency, design time tools, and run-time management.

From these design principles, and taking into consideration detailed use cases and target applications, it is possible to identify requirements that form the basis for a more detailed architecture design. Different sets of requirements have been identified in the already-referenced work, and the reader interested in the more detailed requirements is referred to ETSI TS 102 689 (2013) and IoT-A UNI (2013).

4.3 **An IoT architecture outline**

We have now arrived at a better understanding of the design objectives and principles that capture the main desired characteristics of an M2M or IoT solution, and we have also identified some high-level capabilities that generally are needed. As described above, there is a rather widely accepted view of what a typical M2M solution looks like. However, there is no generally accepted M2M systems architecture or universal set of standards that is widely acknowledged. What is state of the art today is mainly coming from a few standardization bodies that have specified either protocols as systems components, or system and functional architectures for various parts of a complete end-to-end M2M architecture.

When it comes to IoT, there is today not a single widely accepted view of what a typical IoT solution looks like. However, as mentioned, there are a number of research activities on the European level that are converging on defining a reference architecture, and these activities are to be found as projects in the Internet of Things European Research Cluster (IERC 2013). The diversity in the possible and sought-after applications, as well as the diversity in deployment scenarios, together produces a large set of different requirements and constraints.

Attempting to produce a single architecture consequently results in a number of optional and conditional requirements, all depending on the particular problem at hand or application in focus. Nevertheless, the identified key features that are needed when building an M2M or IoT solution can now be put together into a larger context by proposing a single view of the main functional capabilities (see Figure 4.3). This is not a strict and formal functional architecture, but provides a conceptual overview. It also follows the approach of looking at the system capabilities from a layered point of view, including highlighting key functions that go across the layers.

Other approaches that are common in describing an architecture are the software approach and network approach that are more focused on how functions are distributed across a network topology. Here the different proposed functional layers and capabilities provided are discussed. For the sake of brevity, we use IoT as a collective term to include both M2M and IoT.

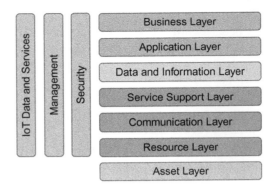

FIGURE 4.3

Functional layers and capabilities of an IoT solution.

At the lowest level is the **Asset Layer**. This layer is, strictly speaking, not providing any functionality within a target solution, but represents the *raison d'être* for any IoT application. The assets of interest are the real-world objects and entities that are subject to being monitored and controlled, as well as having digital representations and identities. The typical examples include vehicles and machinery, fixed infrastructures such as buildings and utility systems, homes, and people themselves — thus being inanimate as well as animate objects. Assets can also be of a more virtual character, being subjective representations of parts of the real world that are of interest to a person or an organization. A typical example of the latter is a set of particular routes used by trucks in a logistics use case. Information of interest may then be traffic intensity, roadwork, or road conditions based on the actual weather situation.

Assets are instrumented with embedded technologies that bridge the digital realm with the physical world, and that provide the capabilities to monitor and control the assets as well as providing identities to the assets. The Resource Layer provides the main functional capabilities of sensing, actuation, and embedded identities. Sensors and actuators in various devices that may be smartphones or Wireless Sensor Actuator Networks (WSANs), M2M devices like smart meters, or other sensor/actuator nodes, deliver these functions. This is also where gateways of different types are placed that can provide aggregation or other capabilities that are closely related to these basic resources. Identification of assets can be provided by different types of tags; for instance, Radio Frequency Identification

(RFID) as in (ISO/IEC RFID 2013), or optical codes like bar codes or Quick Response (QR) codes. The topic of devices and gateways is further dealt with in Section 5.1.

The purpose of the **Communication Layer** is to provide the means for connectivity between the resources on one end and the different computing infrastructures that host and execute service support logic and application logic on the other end. Different types of networks realize the connectivity, and it is customary to differentiate between the notion of a Local Area Network (LAN) and a Wide Area Network (WAN). WANs can be realized by different wired or wireless technologies, for instance, fiber or Digital Subscriber Line (DSL) for the former, and cellular mobile networks, satellite, or microwave links for the latter. WANs can also be provided by different actors, where some networks can be regarded as public (i.e. offered as commercial services for the general public) or as private (i.e. dedicated networks that provide services in a more closed business or entirely company internal environment). Particularly in the mobile network industry, there are different models for how the communications services are provided that include wholesale of access, and dedicated virtual network operators that focus on managed M2M connectivity offerings without owning licensed mobile spectrum or actual network resources. When it comes to LANs, there are many examples of different types, and there is also no stringent definition of what might be considered a LAN or a WAN. Prime examples of LANs include Wireless Personal Area Networks (WPANs; also known as Body Area Networks, BANs) for fitness or healthcare applications, Home or Building Area Networks (HANs and BANs, respectively) used in automation and control applications, and Neighborhood Area Networks (NANs), which are used in the Distribution Grid of a Smart Electricity Grid. Communication can also be used in more *ad hoc* scenarios. Vehicle-to-Vehicle (V2V) is one example that can target safety applications like collision avoidance or car platooning. As opposed to the situation for WANs, the interface technologies used within LANs are characterized by being very industry segment-specific, supported by a plethora of different standards, or even being proprietary or at best de facto standards. LANs use both wired and wireless technologies. General examples of wired LANs include Ethernet and Power Line Communication (PLC), whereas twisted pair (KNX 2013) and (BACnet 2013) over RS-232 are two detailed examples from the building automation industry. Prominent examples of wireless LAN networking technologies include the (IEEE 802.11 2013) and (IEEE 802.15.4 2013) families, as well as (Bluetooth 2013), which has a recent protocol addition called (Bluetooth Low Energy

2013) that targets typical IoT applications. IEEE 802.15.4 is the basis for protocol stacks that target different M2M and IoT applications, for instance, the ZigBee specifications (Zigbee 2013a), the proprietary protocol stack (Z-Wave 2013) for home automation, and ISA100.11a (ISA100 2013) for industry automation. Many of the existing legacy industry-specific LAN protocol stacks do not use IP as the networking protocol, but there is a growing number of examples where the legacy protocol stacks are migrated towards IP, for instance, ZigBee IP, BACnet over IP, and IPv6 over Low power Wireless Personal Area Networks (6LoWPAN) Bluetooth (IETF 6LoWPAN BTLE 2013). To provide an end-to-end communication service that bridges a LAN and a WAN, gateways are used. From a communication layer perspective, gateways are primarily used to do interworking or protocol translation at different levels of the protocol stack. This can involve the physical and link layers, but it can also involve interworking on the communications or messaging level, for example, to do interworking between a legacy protocol like ZigBee to exchanging service operations using HTTP as the means for communication. Section 5.2 deals with the various LAN and WAN aspects and technologies in detail.

As described earlier, IoT applications benefit from simplification by relying on support services that perform common and routine tasks. These support services are provided by the **Service Support Layer** and are typically executing in data centers or server farms inside organizations or in a cloud environment. These support services can provide uniform handling of the underlying devices and networks, thus hiding complexities in the communications and resource layers. Examples include remote device management that can do remote software upgrades, remote diagnostics or recovery, and dynamically reconfigure application processing such as setting event filters. Communication-related functions include selection of communication channels if different networks can be used in parallel, for example, for reliability purposes, and publish—subscribe and message queue mechanisms. Location Based Service (LBS) capabilities and various Geographic Information System (GIS) services are also important for many IoT applications. Of more specific relevance for IoT are services that relate to sensor originating data and actuation services, and services that relate to different tags like RFID. A directory that holds information of available resources and associated service capabilities that can function as a rendezvous mechanism is one example. In such a directory, nodes in WSANs can publish themselves with service descriptions and how to be reached. Applications then perform look-ups to find which device can provide the sensor reading of interest. Another directory service example is the Object Naming Service of EPCGlobal (GS1 EPCGlobal 2013) that can resolve an RFID code to a URL where information about the tagged object can be found. Other repositories can hold information about the persistent real-world entities that are of interest to identify, monitor, and control, such as the Entity Directory from the SENSEI project or the Electronic Product Code Information Services (EPCIS) of EPCGlobal. In general, data storage for anything from raw data to knowledge representations, and processing capabilities such as data and event capture, filtering, and stream processing are different core common services for many IoT applications. Examples of different support services are provided in Chapters 6, 7, and 8.

Where the Resource, Communication, and Service Support layers have concrete realizations in terms of devices and tags, networks and network nodes, and computer servers, the **Data and Information Layer** provides a more abstract set of functions as its main purposes are to capture knowledge and provide advanced control logic support. Key concepts here include data and information models and knowledge representation in general, and the focus is on the organization of information. We refer to a Knowledge Management Framework (KMF) as a collective term to include data, information, domain-specific knowledge, actionable services descriptions as, for example, represented by single actuators or more complex composite sensing and actuation services, service descriptors, rules, process or workflow descriptions, etc. The concept of KMF is further described in Section 5.7. The KMF needs to integrate anything from single pieces of data from individual sensors to highly domain-specific expert knowledge into a common knowledge fabric. Key concepts to construct the KMF include semantic annotation, Linked Data (Bizer et al., 2009), and building different ontologies. Knowledge is highly dynamic, and different techniques are used to capture knowledge as insights, as well as consume knowledge to learn, draw conclusions, propose or even make decisions based on past experiences, current knowledge, and predicted outcomes of certain actions.

The **Application Layer** in turn provides the specific IoT applications. There is an open-ended array of different applications, and typical examples include smart metering in the Smart Grid, vehicle tracking, building automation, or participatory sensing (PS). Part III of this book is devoted to providing examples of different IoT applications.

The final layer in our architecture outline is the **Business Layer**, which focuses on supporting the core business or operations of any enterprise, organization, or individual that is interested in IoT applications. This is where any integration of the IoT applications into business processes and enterprise systems takes place. The enterprise systems can, for example, be Customer Relationship Management (CRM), Enterprise Resource Planning (ERP), or other Business Support Systems (BSS). The business layer also provides exposure to APIs for third parties to get access to data and information, and can also contain support for direct access to applications by human users; for instance, city portal services for citizens in a smart city context, or providing necessary data visualizations to the human workforce in a particular enterprise. The business layer relies on IoT applications as one set of enablers out of many (e.g. field force automation), and takes care of necessary orchestration and composition to support a business process workflow. A detailed discussion on business integration is provided in Section 5.4.

In addition to the functional layers, three functional groups cross the different layers, namely Management, Security, and IoT Data and Services. The former two are well known functions of a system solution, whereas the latter one is more specific to IoT.

Management, as the name implies, deals with management of various parts of the system solution related to its operation, maintenance, administration, and provisioning. This includes management of devices, communications networks, and the general Information Technology (IT) infrastructure as well as configuration and provisioning data, performance of services delivered, etc. M2M management aspects that are covered in Chapters 7 and 8.

Security is about protection of the system, its information and services, from external threats or any other harm. Security measures are usually required across all layers, for instance, providing communication security and information security. Trust and identity management, and authentication and authorization, are key capabilities. From an IoT perspective, management of privacy via, for example, anonymization, is in many instances a specific requirement.

The final functional group of our outlined architecture is denoted **Data and Services**. Data and Service processing can, from a topological perspective, be done in a very distributed fashion and at different levels of complexity. Basic event filtering and simpler aggregation, such as data averaging, can take place in individual sensor nodes in WSANs, contextual metadata such as location and temporal information can be added to sensor readings, and further aggregation can take place higher up in the network topology. More advanced processing is, for instance, data mining and data analytics that can be done in near real-time. This functional group thus represents the vertical flow of data into knowledge, the abstraction of data and services in different levels, and the process steps of extracting knowledge.

As the knowledge layer is focused on the organization and representation of knowledge, this functional group is focused on the different processing steps in the data and services value chain, thus at different levels of granularity and abstraction. Different technologies are used to support the different levels of knowledge extraction, processing, reasoning, and decision-making. Well-known technologies here include stream processing, analytics, machine learning, reasoning, and inferencing. Section 5.6 provides a description of the technologies and tools for data and service processing.

What is not reflected in the architecture outline is the lifecycle aspect of an IoT system solution. Lifecycle aspects of interest include the planning phase for any deployment, and the design phase that involves both systems integration and application development, where APIs and SDKs are important. The actual deployment involving the steps of configuration and provisioning takes place before any solution is put into actual operation. These different steps are outside the scope of this book, but some aspects related to the deployment phase are covered in Chapter 8.

The approach to reach an applied architecture is provided in Chapter 7, whereas Part III of the book provides examples of applied architectures and information on system solutions for selected and typical M2M and IoT applications.

4.4 Standards considerations

The purpose here is not to provide an overview of relevant standards, but to provide an overview of the landscape in which various relevant standards are developed. It is not exhaustive, but will serve as an illustration that the standardization around M2M and IoT is rather complex and multi-dimensional (see Figure 4.4). The primary objective of any technology-oriented standardization activity is to provide a set of agreed-upon specifications that typically address issues like achieving interoperability in a market with many actors and suppliers.

The first consideration is that standards are developed across a number of different industries. There are a number of standardization organizations and bodies, both proper Standards Development Organizations (SDO) as well as special interest groups and alliances that develop standards specifications. Different national and international bodies ratify standards

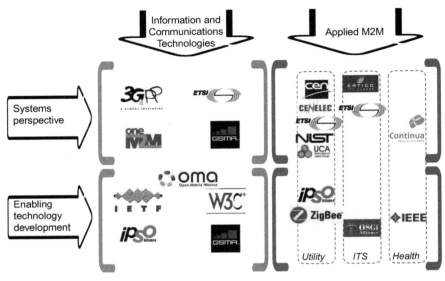

FIGURE 4.4

The landscape of M2M and IoT standardization.

by SDOs, whereas standard specifications developed by special interest groups and alliances are normally agreed-upon and adopted by market actors such as technology manufacturers. Examples of international SDOs are the International Telecommunications Union (ITU) and the International Organization for Standardization (ISO), whereas the European Telecommunications Standards Institute (ETSI) and the European Committee for Electrotechnical Standardization (CENELEC) are examples of regional SDOs. Other independent international standardization organizations include the World Wide Web Consortium (W3C) and the Internet Engineering Task Force (IETF). From an M2M and IoT perspective, one can make a distinction between standards developed within the Information and Communications Technology (ICT) industry, and standards that are developed within a specific industry segment, such as the Health, Transportation, or Electricity industry segments. The ICT industry develops technologies that are targeting use in different other industry segments, and the applied IoT industry segments make use of the ICT standards to varying degrees in developing their standards. As all these industries have a long history of providing their own industry-specific standards, there is an inheritance and legacy of practices and technologies that continue to develop, but as we see more and more converging interests across industries, standards will have to cater to reducing technology fragmentation.

The second consideration is that some standardization activities define entire systems or parts of systems, and other standards organizations target development of specific pieces of technologies, for instance, specific protocols. System standards can address a 3G mobile communication network as defined within the 3rd Generation Partnership Project (3GPP) or standards towards the Smart Grid as done by the National Institute of Standards and Technology (NIST). Organizations like the IETF, on the other hand, focus on developing the protocol suite of the Internet without any effort to specify a system standard beyond what is already in existence in a few key IETF Request For Comments (RFC) such as RFC1958, establishing the Architectural Principles of the Internet. The natural observation is that system standards rely on the enabling technology components as the foundation, but as there generally are many competing technology components (e.g. protocol stacks), the adoption into a system standard is not a straightforward route.

The third and final consideration is about the lifecycle process of standards. Many times, standards are emerging as a result of collaborative research involving both academia and industry. In other situations, technology selection for standardization can happen as part of regulatory or legislative processes. Within the European Union, the European Commission has issued so called Mandates that can have a direct impact on the choice of technology, which hence precedes any subsequent standardization activity. An example of this is the European Mandate M490 (EC M490 2011) on the Smart Grid that was issued by the European Commission to the European Standardization Organizations to come together to develop and update a set of consistent standards within a common European framework that integrates various ICT and electrical architectures and processes to achieve interoperability for the European Smart Grid. As a conclusion, technology selection does not only happen in the process of standardization.

The relevant standards for IoT and M2M are covered in the respective chapters and sections of Parts II and III of the book.

IoT Technologies and Architectures

Part I provided an overview of the vision and market conditions for M2M and the move towards IoT.

In the following chapters, we turn our attention to the technology building blocks that form the basis of M2M and IoT solutions. We then delve into an overview of the architectural foundations of IoT, outlining the European Telecommunications Standards Institute (ETSI) and Internet of Things-Architecture (IoT-A) standards in detail.

Part III then outlines the application of these technologies within real-world use case contexts, illustrating how the M2M/IoT vision combines with the reference architectures to create real-world value.

M2M and IoT Technology Fundamentals

5

CHAPTER OUTLINE

From Machine-to-Machine to the Internet of Things: Introduction to a New Age of Intelligence.
DOI: http://dx.doi.org/10.1016/B978-0-12-407684-6.00005-X

In this chapter, we present an overview of technology fundamentals — or building blocks — that form the basis of M2M and IoT. We cover devices and gateways, local and wide area networking (LAN/WAN), data management, business processes, and cloud and analytics technologies. The next chapters then place these technologies within the architectural frameworks for M2M and IoT.

5.1 Devices and gateways

5.1.1 Introduction

As we discussed in Chapter 1, embedded processing is evolving, not only towards higher capabilities and processing speeds, but also towards allowing the smallest of applications to run on them. There is a growing market for small-scale embedded processing such as 8-, 16-, and 32-bit microcontrollers with on-chip RAM and flash memory, I/O capabilities, and networking interfaces such as IEEE 802.15.4 that are integrated on tiny System-on-a-Chip (SoC) solutions. Such devices enable very constrained devices with a small footprint of a few mm^2 and with a very low power consumption in the milli- to micro-Watt range, but which are capable of hosting an entire Transmission Control Protocol/Internet Protocol (TCP/IP) stack, including a small web server.

There is a full spectrum of M2M and IoT devices, and to avoid confusion, let's start with explaining what is referred to as a device here: A device is a hardware unit that can sense aspects of it's environment and/or actuate, i.e. perform tasks in its environment.

A device can be characterized as having several properties, including:

- **Microcontroller**: 8-, 16-, or 32-bit working memory and storage.
- **Power Source**: Fixed, battery, energy harvesting, or hybrid.
- **Sensors and Actuators**: Onboard sensors and actuators, or circuitry that allows them to be connected, sampled, conditioned, and controlled.
- **Communication**: Cellular, wireless, or wired for LAN and WAN communication.
- **Operating System (OS)**: Main-loop, event-based, real-time, or full-featured OS.
- **Applications**: Simple sensor sampling or more advanced applications.
- **User Interface**: Display, buttons, or other functions for user interaction.
- **Device Management (DM)**: Provisioning, firmware, bootstrapping, and monitoring.
- **Execution Environment (EE)**: Application lifecycle management and Application Programming Interface (API).

For several reasons, one or more of these functions are often hosted on a gateway instead. This can be to save battery power, for example, by letting the gateway handle heavy functions such as WAN connectivity and application logic that requires a powerful processor. This also leads to reduced costs because these are expensive components.

Another reason is to reduce complexity by letting a central node (the gateway) handle functionality such as device management and advanced applications, while letting the devices focus on sensing and actuating.

Table 5.1 Example Characteristics of the Device Types

	CPU	Memory	Power	Comm	OS, EE
Basic	8-bit PIC, 8-bit 8051, 32-bit Cortex-M	Kilobytes	Battery	802.15.4, 802.11, Z-Wave	Main-loop, Contiki, RTOS[a]
Advanced	32-bit ARM9, Intel Atom	Megabytes	Fixed	802.11, LTE, 3G, GPRS	Linux, Java, Python

[a]*Real-time operating system.*

5.1.1.1 Device types

There are no clear criteria today for categorizing devices, but instead there is more of a sliding scale. Within the book, we group devices into two categories (Table 5.1):

- **Basic Devices**: Devices that only provide the basic services of sensor readings and/or actuation tasks, and in some cases limited support for user interaction. LAN communication is supported via wired or wireless technology, thus a gateway is needed to provide the WAN connection.
- **Advanced Devices**: In this case the devices also host the application logic and a WAN connection. They may also feature device management and an execution environment for hosting multiple applications. Gateway devices are most likely to fall into this category.

5.1.1.2 Deployment scenarios for devices

Deployment can differ for basic and advanced deployment scenarios. Example deployment scenarios for basic devices include:

- **Home Alarms**: Such devices typically include motion detectors, magnetic sensors, and smoke detectors. A central unit takes care of the application logic that calls security and sounds an alarm if a sensor is activated when the alarm is armed. The central unit also handles the WAN connection towards the alarm central. These systems are currently often based on proprietary radio protocols.
- **Smart Meters**: The meters are installed in the households and measure consumption of, for example, electricity and gas. A concentrator gateway collects data from the meters, performs aggregation, and periodically transmits the aggregated data to an application server over a cellular connection. By using a capillary network technology

(e.g. 802.15.4), it's possible to extend the range of the concentrator gateway by allowing meters in the periphery to use other meters as extenders, and interface with handheld devices on the Home Area Network side.

- **Building Automation Systems** (BASs): Such devices include thermostats, fans, motion detectors, and boilers, which are controlled by local facilities, but can also be remotely operated.
- **Standalone Smart Thermostats**: These use Wi-Fi to communicate with web services.

Examples for advanced devices, meanwhile, include:

- **Onboard units in cars** that perform remote monitoring and configuration over a cellular connection.
- **Robots and autonomous vehicles** such as unmanned aerial vehicles that can work both autonomously or by remote control using a cellular connection.
- **Video cameras** for remote monitoring over 3G and LTE.
- **Oil well monitoring** and collection of data points from remote devices.
- **Connected printers** that can be upgraded and serviced remotely.

The devices and gateways of today often use legacy technologies such as KNX, Z-Wave, and ZigBee, but the vision for the future is that every device can have an IP address and be (in)directly connected to the Internet.

Some of the examples listed above (e.g. the BAS) require some form of autonomous mode, where the system operates even without a WAN connection. Also, in these cases it's possible to use IoT technologies to form an "Intranet of Things."

5.1.2 Basic devices

These devices are often intended for a single purpose, such as measuring air pressure or closing a valve. In some cases several functions are deployed on the same device, such as monitoring humidity, temperature, and light level.

The requirements on hardware are low, both in terms of processing power and memory. The main focus is on keeping the bill of materials (BOM) as low as possible by using inexpensive microcontrollers with built-in memory and storage, often on an SoC-integrated circuit with all main components on one single chip (Figure 5.1). Another common goal is to enable battery as a power source, with a lifespan of a year and upwards by using ultra-low energy microcontrollers.

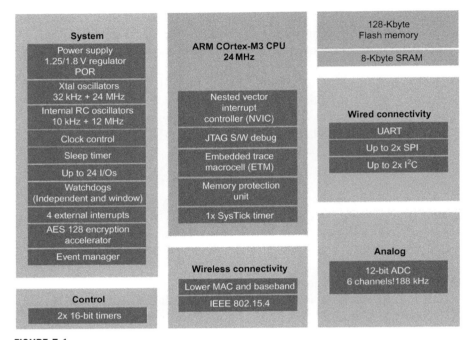

FIGURE 5.1

Example of a microcontroller with integrated STM32W-RFCKIT.

(STMicroelectronics 2013)

The microcontroller typically hosts a number of ports that allow integration with sensors and actuators, such as General Purpose I/O (GPIO) and an analog-to-digital converter (ADC) for supporting analog input. For certain actuators, such as motors, pulse-width modulation (PWM) can be used.

As low-power operation is paramount to battery-powered devices, the microcontroller hosts functions that facilitate sleeping, such as interrupts that can wake up the device on external and internal events, e.g. when there is activity on a GPIO port or the radio, as well as timer-based wake ups. Some devices even go as far as harvesting energy from their environment, e.g. in the form of solar, thermal, and physical energy.

To interact with peripherals such as storage or display, it's common to use a serial interface such as SPI, I²C, or UART. These interfaces can also be used to communicate with another microcontroller on the device. This is common when the there is a need for offloading certain tasks, or when in some cases the entire application logic is put on a separate host processor.

It's not unusual for the microcontroller to also contain a security processor, e.g. to accelerate Advanced Encryption Standard (AES). This is necessary to allow encrypted communication over the radio link without the need for a host processor.

Because a basic device lacks a WAN interface according to our definition, a gateway of some form is necessary. The gateway together with the connected devices form a capillary network. The microcontroller contains most of the radio functions needed for communicating with the gateway and other devices in the same capillary network. An external antenna is, however, necessary, and preferably a filter that removes unwanted frequencies, e.g. a surface acoustic wave (SAW) filter.

Due to limited computational resources, these devices commonly do not use a typical OS. It may be something as simple as a single-threaded main-loop or a low-end OS such as FreeRTOS, Atomthreads, AVIX-RT, ChibiOS/RT, ERIKA Enterprise, TinyOS, or Thingsquare Mist/Contiki. These OSes offer basic functionality, e.g. memory and concurrency model management, (sensor and radio) drivers, threading, TCP/IP, and higher-level protocol stacks.

The actual application logic is located on top of the OS or in the main-loop. A typical task for the application logic is to read values from the sensors and to provide these over the LAN interface in a semantically correct manner with the correct units.

For this class of devices, the constrained hardware and non-standard software limit third-party development and make development quite cost-intensive.

5.1.3 Gateways

A gateway serves as a translator between different protocols, e.g. between IEEE 802.15.4 or IEEE 802.11, to Ethernet or cellular.

There are many different types of gateways, which can work on different levels in the protocol layers. Most often a gateway refers to a device that performs translation of the physical and link layer, but application layer gateways (ALGs) are also common. The latter is preferably avoided because it adds complexity and is a common source of error in deployments.

Some examples of ALGs include the ZigBee Gateway Device (ZigBee Alliance 2011), which translates from ZigBee to SOAP and IP, or gateways that translate from Constrained Application Protocol (CoAP) to HyperText Transfer Protocol/Representational State Transfer (HTTP/REST).

For some LAN technologies, such as 802.11 and Z-Wave, the gateway is used for inclusion and exclusion of devices. This typically works by activating the gateway into inclusion or exclusion mode and by pressing a button on the device to be added or removed from the network. We cover network technologies in more detail in Section 5.2: Local and wide area networking.

For very basic gateways, the hardware is typically focused on simplicity and low cost, but frequently the gateway device is also used for many other tasks, such as data management, device management, and local applications. In these cases, more powerful hardware with GNU/Linux is commonly used. The following sections describe these additional tasks in more detail.

5.1.3.1 Data management

Typical functions for data management include performing sensor readings and caching this data, as well as filtering, concentrating, and aggregating the data before transmitting it to back-end servers. Data Management is covered in Section 5.3.

5.1.3.2 Local applications

Examples of local applications that can be hosted on a gateway include closed loops, home alarm logic, and ventilation control, or the data management function above and in Section 5.3. The benefit of hosting this logic on the gateway instead of in the network is to avoid downtime in case of WAN connection failure, minimize usage of costly cellular data, and reduce latency.

To facilitate efficient management of applications on the gateway, it's necessary to include an execution environment. The execution environment is responsible for the lifecycle management of the applications, including installation, pausing, stopping, configuration, and uninstallation of the applications. A common example of an execution environment for embedded environments is OSGi, which is based on Java: applications are built as one or more Bundles, which are packaged as Java JAR files and installed using a so-called Management Agent. The Management Agent can be controlled from, for example, a terminal shell or via a protocol such as CPE WAN Management Protocol (CWMP).

Bundle packages can be retrieved from the local file system or over HTTP, for example. OSGi also provides security and versioning for Bundles, which means that communication between Bundles is controlled, and several versions of them can exist. The benefit of versioning and the lifecycle management functions is that the OSGi environment never needs to be shut down when upgrading, thus avoiding downtime in the system. Also, Linux can be used as an execution environment.

5.1.3.3 Device management

Device management (DM) is an essential part of the IoT and provides efficient means to perform many of the management tasks for devices:

- **Provisioning**: Initialization (or activation) of devices in regards to configuration and features to be enabled.
- **Device Configuration**: Management of device settings and parameters.
- **Software Upgrades**: Installation of firmware, system software, and applications on the device.
- **Fault Management**: Enables error reporting and access to device status.

Examples of device management standards include TR-069 and OMA-DM.

In the simplest deployment, the devices communicate directly with the DM server. This is, however, not always optimal or even possible due to network or protocol constraints, e.g. due to a firewall or mismatching protocols. In these cases, the gateway functions as mediator between the server and the devices, and can operate in three different ways:

- If the devices are visible to the DM server, the gateway can simply forward the messages between the device and the server and is not a visible participant in the session.
- In case the devices are not visible but understand the DM protocol in use, the gateway can act as a proxy, essentially acting as a DM server towards the device and a DM client towards the server.
- For deployments where the devices use a different DM protocol from the server, the gateway can represent the devices and translate between the different protocols (e.g. TR-069, OMA-DM, or CoAP). The devices can be represented either as virtual devices or as part of the gateway (which is typically also a device that is managed by the DM server).

5.1.4 Advanced devices

As mentioned earlier, the distinction between basic devices, gateways, and advanced devices is not cut in stone, but some features that can characterize an advanced device are the following:

- A powerful CPU or microcontroller with enough memory and storage to host advanced applications, such as a printer offering functions for copying, faxing, printing, and remote management.
- A more advanced user interface with, for example, display and advanced user input in the form of a keypad or touch screen.
- Video or other high bandwidth functions.

It's not unusual for the advanced device to also function as a gateway for local devices on the same LAN.

For these more computionally capable devices, the OS can be, for example, GNU/Linux or a commerical RTOS, such as ENEA's OSE, WindRiver's VxWorks, or Blackberry's QNX. This class of devices comes with optimized and high-performance IP stacks, thus making networking a non-issue.

By offering a more common and open OS, along with community-standardized APIs, software libraries, programming languages, and development tools, the number of potential developers grows significantly.

5.1.5 Summary and vision

This section covered different device classes and the role of the gateway in an M2M or IoT deployment. There are also other aspects that must be taken into account in regards to devices. The most important of these is security, both in terms of physical security as well as software and network security. As this is a very extensive topic, it is out of the scope of this book.

Another aspect that needs to be managed is the matter of external factors that can affect the operation of the devices, such as rain, wind, chemicals, and electromagnetic influences. These elements are slowly being understood and studied in the context of cyber-physical systems, as IoT applications move from the laboratory into real-world deployments. Fundamentally, these external factors necessitate adaptability and situational awareness capabilities as features of the devices in the field, which are typically unaccounted for during the software engineering phase of development.

One of the major effects that the IoT will have on devices is to disrupt the current value chains, where one actor controls everything from device to service. This will happen due to standardization and consolidation of technologies, such as protocols, OSes, software and programming languages (e.g. Java for embedded devices), and the business drivers discussed in Chapter 2 . New types of actors will be able to enter the market, e.g. specialized device vendors, cloud solution providers, and service providers. Standardization will improve interoperability between devices, as well as between devices and services, resulting in commoditization of both.

Another expected outcome of improved interoperability is the possibility to reuse the same device for multiple services; for example, a motion detector can be used both for security purposes as well as for reducing energy consumption by detecting when no one is in the room.

The barrier for new developers will further be reduced thanks to the consolidation of software and interfaces, e.g. it will be possible to interact with a device using simple HTTP/REST and to easily install a Java application on a device, resulting in an increased number of developers.

Thanks to developments in hardware and network technologies, entirely new device classes and features are expected, such as:

- Battery-powered devices with ultra-low power cellular connections.
- Devices that harvest energy from their environment.
- Smart bandwidth management and protocol switching, i.e. using adaptive RF mechanisms to swap between, for example, Bluetooth LE and IEEE 802.15.4.
- Multi-radio/multi-rate to switch between bands or bit rates (slower bit rate implies better sensitivity at longer range).
- Microcontrollers with multicore processors.
- Novel software architectures for better handling of concurrency.
- The possibility to automate the design of integrated circuits based on business-level logic and use case.

All these improvements that the IoT brings will remove the final barriers that have been holding back the market for M2M.

In the next section, we cover LANs and WANs — the technology building blocks that allow devices to communicate with Information and Communications Technology (ICT) systems and the wider world.

5.2 Local and wide area networking

5.2.1 The need for networking

A network is created when two or more computing devices exchange data or information. The ability to exchange pieces of information using telecommunications technologies has changed the world, and will continue to do so for the foreseeable future, with applications emerging in nearly all contexts of contemporary and future living. Typically, devices are known as "nodes" of the network, and they communicate over "links."

In modern computing, nodes range from personal computers, servers, and dedicated packet switching hardware, to smartphones, games consoles, television sets and, increasingly, heterogeneous devices that are generally characterized by limited resources and functionalities. Limitations typically include computation, energy, memory, communication (range, bandwidth, reliability, etc.) and application specificity (e.g. specific sensors,

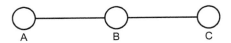

FIGURE 5.2

A network.

actuators, tasks), etc. Such devices are typically dedicated to specific tasks, such as sensing, monitoring, and control (discussed later).

Network links rely upon a physical medium, such as electrical wires, air, and optical fibers, over which data can be sent from one network node to the next. It is not uncommon for these media to be grouped either as wired or wireless.

A selected physical medium determines a number of technical and economic considerations. Technically, the medium selected, or more accurately, the technological solution designed and implemented to communicate over that medium, is the primary enabler of bandwidth — without which, certain applications are infeasible. Simultaneously, different technological solutions require certain economic considerations, such as the cost of deployment and maintenance of the networking infrastructure. For example, consider the cost of embedding wires across a metropolitan, or larger, geographic region (e.g. electricity and legacy telephone networks).

When direct communication between two nodes over a physical medium is not possible, networking can allow for these devices to communicate over a number of hops. In order to achieve this, nodes of the network must have an awareness of all nodes in the network with which they can indirectly communicate. This can be a direct connection over one link (edge, the transition or communication between two nodes over a link), or knowledge of a route to the desired (destination) node by communicating through cooperating nodes, over multiple edges. Consider Figure 5.2.

This is the simplest form of network that requires knowledge of a route to communicate between nodes that do not have direct physical links. Therefore, if node A wishes to transfer data to node C, it must do so through node B. Thus, node B must be capable of the following: communicating with both node A and node C, and advertising to node A and node C that it can act as an intermediary.

Basic networking requirements have become explicit. It is essential to uniquely identify each node in the network, and it is necessary to have cooperating nodes capable of linking nodes between which physical links do not exist. In modern computing, this equates to IP addresses and routing tables. Thanks to standardization, IP in particular, physical media (links) no longer need to be the same across the network, and the nodes need not have the same capabilities and/or mission.

Beyond the basic ability to transfer data, the speed and accuracy with which data can be transferred is of critical importance to the application. Irrespective of the ability to link devices, without the necessary bandwidth, some applications are rendered impossible. Consider the differences between streaming video from a surveillance camera, for example, and an intrusion-detection system based on a passive sensor. Simplistically, streaming video requires high bandwidth, whereas transmitting a small amount of information about the detection of an intruder requires a tiny amount of bandwidth, but a higher degree of reliability with respect to both the communications link and the accuracy of the detection.

Today, we have complex, heterogeneous networks. The simple example above is useful for explaining the basics of networking to a child, but is also useful when abstracting the types of nodes that A, B, and C might be, the different physical links between them, and their methods of interaction.

Abstracting, consider that node A is a device that can only communicate over a particular wireless channel of limited range (e.g. Channel 11 in the 2.4 GHz ISM band, over less than 200 meters). Node B is capable of communicating with node A, but also with an application server with service capabilities (node C, with which it can connect using wired Ethernet, e.g. over a complex link using a standardized protocol and/or web service such as REST at the application layer) over the Internet. Now consider that node B may be connected to a sub-network (of child nodes, similar to node A) of up to thousands of similarly constrained devices ($A_1 \ldots A_n$).

These thousands of devices may be equipped with sensors, deployed specifically to monitor some physical phenomenon. They can only communicate with one another and node B, and may communicate with each other over single or multiple hops (thus increasing range of the sensing field, not all nodes requiring direct connectivity with B). This is representative of a traditional Wireless Sensor Network (WSN).

Consider that the owner of the WSN wishes to obtain the data from each of the ($A_1 \ldots A_n$) devices in the WSN. However, the preferred way to read the data is through a web browser, or application on a smartphone/tablet, via node C. Therefore, a networking solution is required to transfer all of the WSN data from nodes $A_1 \ldots A_n$ to node C, through node B. This is now a complex networking infrastructure, and is representative of many potential embodiments of M2M and IoT technologies. This concept maps directly to the M2M Functional Architecture, where nodes $A_1 \ldots A_n$ are an

M2M Area Network, node B is an M2M Gateway, and node C is representative of M2M Service Capabilities and Applications.

A **Local Area Network** (LAN) was traditionally distinguishable from a **Wide Area Network** (WAN) based on the geographic coverage requirements of the network, and the need for third party, or leased, communication infrastructure. In the case of the LAN, a smaller geographic region is covered, such as a commercial building, an office block, or a home, and does not require any leased communications infrastructure.

WANs provide communication links that cover longer distances, such as across metropolitan, regional, or by textbook definition, global geographic areas. In practice, WANs are often used to link LANs and Metropolitan Area Networks (MAN) — where LAN technologies cannot provide the communications ranges to otherwise interconnect — and commonly to link LANs and devices (including smart phones, Wi-Fi routers that support LANs, tablets, and M2M devices) to the Internet. Quantitatively, LANs tended to cover distances of tens to hundreds of meters, whereas WAN links spanned tens to hundreds of kilometers.

There are differences between the technologies that enable LANs and WANs. In the simplest case for each, these can be grouped as wired or wireless. The most popular wired LAN technology is Ethernet. Wi-Fi is the most prevalent wireless LAN (WLAN) technology.

Wireless WAN (WWAN), as a descriptor, covers cellular mobile telecommunication networks, a significant departure from WLAN in terms of technology, coverage, network infrastructure, and architecture. The current generation of WWAN technology includes LTE (or 4G) and WiMAX.

Acting as a link between LANs and Wireless Personal Area Networks (WPANs), M2M Gateway Devices (see Section 5.1) typically include cellular transceivers, and allow seamless IP-connectivity over heterogeneous physical media. An example of an M2M Gateway's logical functionality is presented in Figure 6.9.

A more intuitive example of a similar device is the wireless access point commonly found in homes and offices. In the home, the "wireless router" typically behaves as a link between the Wi-Fi (WLAN, and thus connected laptops, tablets, smartphones, etc. commonly found in the home) and Digital Subscriber Line (DSL) broadband connectivity, traditionally arriving over telephone lines. "DSL" refers to Internet access carried over legacy (wired) telephone networks, and encompasses numerous standards and variants. "Broadband" indicates the ability to carry multiple signals over a number of frequencies, with a typical minimum bandwidth of 256 kbps. In the office, the Wi-Fi wireless access points are typically connected to the wired corporate (Ethernet) LAN, which is subsequently connected to a wider area network and Internet backbone, typically provided by an Internet Service Provider (ISP).

Considering M2M and IoT applications, there are likely to exist a combination of traditional networking approaches. The need exists to interconnect devices (generally integrated microsystems) with central data processing and decision support systems, in addition to one another. The business logic and requirements for each embodiment will differ on a case-by-case basis.

Practically, these devices will not warrant individual connections to leased networking infrastructure (e.g. putting a SIM card in each device and using the cellular network for fast IP connectivity). This approach is thought to be prohibitive due to cost, among other factors. A more likely scenario is where, similar to WLAN technologies, a geographic region can be covered by a network of devices that connect to the Internet via a gateway device, which may use a leased network connection.

The potential complexity of these networks is enormous. For example, a gateway device can access the IP backbone over a WWAN (e.g. GPRS/UMTS/LTE/WiMAX) link, or over a WLAN link, where the leased infrastructure would be that of the ISP providing backbone connectivity to the WLAN in its own right, as above.

It is worth extending the consideration of WAN and LAN to encompass the idea of WPANs, which is the description used for the newer standards that govern low-power, low-rate networks suitable for M2M and IoT applications. Indeed, the standard upon which many popular recent networking technologies are built (including ZigBee, WirelessHART, Isa100.11a, and other IETF initiatives such as 6LoWPAN, RPL, and CoAP, all discussed later in this chapter) is "IEEE 802.15.4 − Wireless Medium Access Control (MAC) and Physical Layer (PHY) Specifications for Low-Rate Wireless Personal Area Networks (LR-WPANs)." This standard was first approved in 2003, and has been subject to numerous amendments over the past decade. These amendments have related to modifying and/or extending PHY parameters to ensure global utility with regard to licensing and application suitability, and modifications to the MAC layer. This is similar to the evolution of Wi-Fi WLAN technology (e.g. IEEE 802.11, a, b, g, n, etc.). The naming convention is ultimately non-intuitive, as communication ranges for IEEE 802.15.4 technology may range from tens of meters to kilometers. It is probably more useful to think of these technologies as "low-rate, low-power" networks.

It is reasonable to suggest that the traditional boundaries between LAN and WAN technologies, and their working definitions, require updating to account for contemporary amendments to the standards and use cases to which they are applicable.

From an M2M perspective, ETSI considers, as part of its functional architecture, M2M Area Networks (the terms "M2M Area Network," "M2M Device Domain," and "capillary networks" are often used interchangeably to describe peripheral networks below the Access and Core Networks). Devices in an M2M Area Network connect to the IP backbone, or Network Domain, via an M2M Gateway device. Typically, a Gateway device is equipped with a cellular transceiver that is physically compatible with UMTS or LTE-Advanced, for example, WWAN. The same device will also be equipped with the necessary transceiver to communicate on the same physical medium as the M2M Area Network(s) in the M2M Device Domain. This is covered in more detail in Chapter 6.

M2M Area Networks may include a plethora of wired or wireless technologies, including: Bluetooth LE/Smart, IEEE 802.15.4 (LR-WPAN; e.g. ZigBee, IETF 6LoWPAN, RPL, CoAP, ISA100.11a, WirelessHART, etc.), M-BUS, Wireless M-BUS, KNX, and Power Line Communication (PLC).

The "Internet of Things," as a term, originated from Radio Frequency Identification (RFID) research, wherein the original IoT concept was that any RFID-tagged "thing" could have a virtual presence on the "Internet." In reality, there is little conceptual dissimilarity between RFID and bar codes, or more recently, QR codes — they simply use different technological means to achieve the same result (i.e. an "object" has an online presence).

The original concept has evolved from a reasonably simple idea, with immediate utility in logistics (i.e. track and trace, inventory management applications), to complex networks, functionalities, and interactions, without any satisfactory working definition(s). As M2M applications become more synonymous with IoT, it is necessary to understand the technologies, limitations, and implications of the networking infrastructure. Essentially, the ability to remotely communicate with devices, and resultant new capabilities, is what sets modern IoT thinking apart from the original concept.

The following sub-sections describe the technologies traditionally used to achieve WAN and LAN. We are further motivated to shift away from conventional thinking on how to describe networking and communications technologies based on simplistic concepts around geographic coverage or leased infrastructure.

5.2.2 Wide area networking

WANs are typically required to bridge the M2M Device Domain to the backhaul network, thus providing a proxy that allows information (data, commands, etc.) to traverse heterogeneous networks. This is seen as a core requirement to provide communications services between the M2M service enablement and the physical deployments of devices in the field. Thus, the WAN is capable of providing the bi-directional communications links between services and devices. This, however, must be achieved by means of physical and logical proxy.

The proxy is achieved using an M2M Gateway Device. Depending on the situation, there are, in general, a number of candidate technologies to select from. As before, the M2M Gateway Device is typically an integrated microsystem with multiple communications interfaces and computational capabilities. It is a critical component in the functional architecture, as it must be capable of handling all of the necessary interfacing to the M2M Service Capabilities and Management Functions, which are mostly described in Chapter 6 and the ETSI releases.

By way of example, consider a device that incorporates both an IEEE 802.15.4-compliant transceiver (a popular legacy example of which is the Texas Instruments CC2420), capable of communicating with a capillary network of similarly equipped devices, and a cellular transceiver (a popular example of which is a Telit UC864-g), that connects to the Internet using the UMTS network. This assumes the handover to the backbone IP network is handled according to the 3GPP specifications. Transceivers (sometimes referred to as modems) are typically available as hardware modules with which the central intelligence of the device (gateway or cell phone) interacts by means of standardized (sometimes vendor-specific) AT Commands.

This device is now capable of acting as a physical proxy between the LR-WPAN, or M2M Device Domain, and the M2M Network Domain. The latest ETSI M2M Functional Architecture is illustrated in Figure 5.3. Device types were discussed in more detail in Section 5.1.

The Access and Core Network in the ETSI M2M Functional Architecture are foreseen to be operated by a Mobile Network Operator (MNO), and can be thought of simply as the "WAN" for the purposes of interconnecting devices and backhaul networks (Internet), thus, M2M Applications, Service Capabilities, Management Functions, and Network Management Functions.

The WAN covers larger geographic regions using wireless (licensed and un-licensed spectra) as well as wire-based access. WAN technologies include cellular networks (using several generations of technologies), DSL, WiMAX, Wi-Fi, Ethernet, Satellite, and so forth.

The WAN delivers a packet-based service using IP as default. However, circuit-based services can also be used in certain situations.

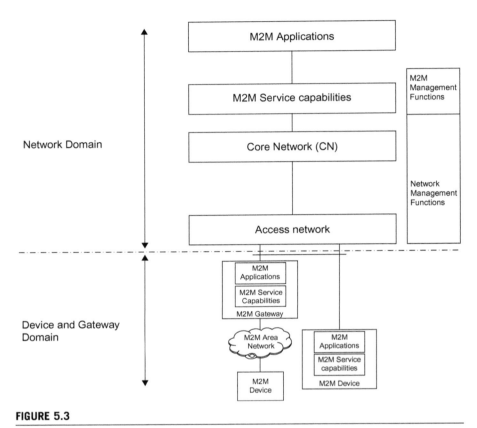

FIGURE 5.3

ETSI M2M Functional Architecture.

In the M2M context, important functions of the WAN include:

- The main function of the WAN is to establish connectivity between capillary networks, hosting sensors, and actuators, and the M2M service enablement. The default connectivity mode is packet-based using the IP family of technologies.

  Many different types of messages can be sent and received. These include messages originating as, for example, a message sent from a sensor in an M2M Area Network and resulting in an SMS received from the M2M Gateway or Application (e.g. by a relevant stakeholder with SMS notifications configured for when sensor readings breach particular sensing thresholds.).

- Use of identity management techniques (primarily of M2M devices) in cellular and non-cellular domains to grant right-of-use of the

WAN resource. The following techniques are used for these purposes:

- MCIM (Machine Communications Identity Module) for remote provisioning of SIM targeting M2M devices.
- xSIM (x-Subscription Identity Module), like SIM, USIM, ISIM.
- Interface identifiers, an example of which is the MAC address of the device, typically stored in hardware.
- Authentication/registration type of functions (device focused).
- Authentication, Authorization, and Accounting (AAA), such as RADIUS services.
- Dynamic Host Configuration Protocol (DHCP), e.g. employing deployment-specific configuration parameters specified by device, user, or application-specific parameters residing in a directory.
- Subscription services (device-focused).
- Directory services, e.g. containing user profiles and various device (s) parameter(s), setting(s), and combinations thereof. M2M-specific considerations include, in particular:
- MCIM (cf. 3GPP SA3 work).
- User Data Management (e.g. subscription management).
- Network optimizations (cf. 3GPP SA2 work).

There may be many suppliers of WAN functionality in a complete M2M solution. It follows that an important function in the M2M Service Enablement domain will be to manage westbound business-to-business (B2B) relations between a number of WAN service providers.

5.2.2.1 3rd generation partnership project technologies and machine type communications

Machine Type Communications (MTC) is heavily referred to in the ETSI documentation. MTC, however, lacks a firm definition, and is explained using a series of use cases. Generally speaking, MTC refers to small amounts of data that are communicated between machines (devices to back-end services and vice versa) without the need for any human intervention. In the 3rd Generation Partnership Project (3GPP), MTC is used to refer to all M2M communication (Jain et al. 2012). Thus, they are interchangeable terms.

5.2.3 Local area networking

Capillary networks are typically autonomous, self-contained systems of M2M devices that may be connected to the cloud via an appropriate

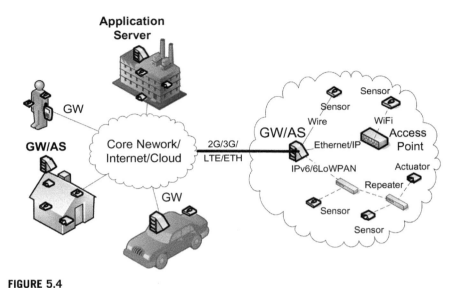

FIGURE 5.4

Capillary networks and their inside view.

Gateway. They are often deployed in controlled environments such as vehicles, buildings, apartments, factories, bodies, etc. (Figure 5.4) in order to collect sensor measurements, generate events should sensing thresholds be breached, and sometimes control specific features of interest (e.g. heart rate of a patient, environmental data on a factory floor, car speed, air conditioning appliances, etc.). There will exist numerous capillary networks that will employ short-range wired and wireless communication and networking technologies.

For certain application areas, there is a need for autonomous local operation of the capillary network. That is, not everything needs to be sent to, or potentially be controlled via, the cloud.

In the event that application-level logic is enforceable via the cloud, some will still need to be managed locally. The complexity of the local application logic varies by application. For example, a building automation network may need local control loop functionality for autonomous operation, but can rely on external communication for configuration of control schemas and parameters.

The M2M devices in a capillary network are typically thought to be low-capability nodes (e.g. battery operated, with limited security capabilities) for cost reasons, and should operate autonomously. For this reason, a GW/application server will naturally also be part of the architected solution for capillary networks.

More and more (currently closed) capillary networks will open up for integration with the enterprise back end systems. For capillary networks that expose devices to the cloud/Internet, IP is envisioned to be the common waist. IPv6 will be the protocol of choice for M2M devices that operate a 6LoWPAN-based stack. IPv4 will still be used for capillary networks operating in non-6LoWPAN IP stacks (e.g. Wi-Fi capillary networks).

In terms of short-range communication technology convergence, an IPv6 stack with 6LoWPAN running above the physical medium is expected. The physical medium may be IEEE 802.15.4 (i.e. wireless), but can also be various PLC or other wired solutions (e.g. Homeplug or P1902).

Legacy ZigBee application profiles will be used in the future in addition to newer ZigBee IP and IEEE 802.15.4 6LoWPAN/RPL/CoAP networks. It is expected that the binary versions of the application profiles will be used for efficiency reasons (e.g. an automation profile device may be a temperature sensor not necessarily connected to the mains).

Market conditions dictate that in 2016, there will continue to be ZigBee and BT LE devices competing for share of the same market. There are no indications that KNX and ZigBee will be consolidated, as there are big players that support these for the smart grid application area. The situation is further exacerbated by the development of the IEEE 802.15.4g standard, a physical layer amendment to support Smart Utility Networks (SUN) — smart grid in particular — designed to operate over much larger geographic distances (wireless links spanning tens of kilometers), and specifically designed for minimal infrastructure, low power, many-device networks.

5.2.3.1 Deployment considerations

The nature of the intended application plays a significant role in determining the appropriate technological solution. Typically, these are defined by the business logics that motivate initial deployment. There are increasing numbers of innovative IoT applications (hardware and software) marketed as consumer products. These range from intelligent thermostats for effectively managing comfort and energy use in the home, to precision gardening tools (sampling weather conditions, soil moisture, etc.). At scale, similar solutions are, and will continue to be, applied in and across industry.

Scaling up for industrial applications and moving from laboratories into the real world creates significant challenges that are not yet fully understood. Low-rate, low-power communications technologies are known to be "lossy." The reasons for this are numerous. They can relate to environmental factors, which impact upon radio performance (such as time-varying stochastic wireless propagation characteristics), technical factors such as performance trade-offs based on the characteristics of medium access control and routing protocols, and physical limitations of devices (including software architectures, runtime and execution environments, computational capabilities, energy availability, local storage, and so forth), and practical factors such as maintenance opportunities (scheduled, remote, accessibility, etc.). Practical constraints for deployment are considered in more detail in Section 7.3.

Numerous deployment environments (factories, buildings, roads, vehicles) are expected in addition to wildly varying application scenarios and operational and functional requirements of the systems. ETSI describes a set of use cases, namely eHealth, Connected Consumer, Automotive, Smart Grid, and Smart Meter, that only capture some of the breadth of potential deployment scenarios and environments that are possible.

Section 5.1 describes the various hardware technologies that comprise the devices and gateways that make up the current and future art in IoT technologies. Notwithstanding, there continues to be fragmentation at the physical layer in terms of communications technologies. Assuming that IP connectivity can be the fundamental mechanism to bridge heterogeneous physical and link layer technologies, it stands to reason that fragmentation can continue such that appropriate technologies are available for the breadth of potential application scenarios.

5.2.3.2 Key technologies
This section details a number of the standards and technologies currently in use and under development that enable *ad hoc* connectivity between the devices that will form the basis of the IoT. These are the communications technologies that are considered to be critical to the realization of massively distributed M2M applications and the IoT at large.

Power Line Communication (PLC) refers to communicating over power (or phone, coax, etc.) lines. This amounts to pulsing, with various degrees of power and frequency, the electrical lines used for power distribution. PLC comes in numerous flavors. At low frequencies (tens to hundreds of Hertz) it is possible to communicate over kilometers with low bit rates (hundreds of bits per second). Typically, this type of communication was used for remote metering, and was seen as potentially useful for the smart grid. Enhancements to allow higher bit rates have led to the possibility of delivering broadband connectivity over power lines. There have been a number of attempts to standardize PLC in recent years. NIST recently included IEEE 1901 (in 2011) and ITU-T G.hn (G.9960-PHY in 2009, and G.9961-Data Link Layer in 2010) as standards for further review for potential use in the smart grid in the United States. In 2011, the ITU referenced G.9903 that specifies the use of IPv6 over PLC, borrowing techniques (specifically 6LoWPAN, below) originally developed in the wireless community.

LAN (and WLAN) continues to be important technology for M2M and IoT applications. This is due to the high bandwidth, reliability, and legacy of the technologies. Where power is not a limiting factor, and high bandwidth is required, devices may connect seamlessly to the Internet via Ethernet (IEEE 802.3) or Wi-Fi (IEEE 802.11). The utility of existing (W) LAN infrastructure is evident in a number of early IoT applications targeted at the consumer market, particularly where integration and control with smartphones is required (irrespective of the actual technical architecture or optimality of the solution).

The IEEE 802.11 (Wi-Fi) standards continue to evolve in various directions to improve certain operational characteristics depending on usage scenario. A widely adopted recent release was IEEE 802.11n, which was specifically designed to enhance throughput (typically useful for streaming multimedia). Ongoing work such as IEEE 802.11ac is developing an even higher throughput version to replace this, focusing efforts in the 5 GHz band.

IEEE 802.11ah is working on an evolution of the 2007 standard that will allow a number of networked devices to cooperate in the <1 GHz (ISM) band. The idea is to exploit collaboration (relaying, or networking in other words) to extend range, and improve energy efficiency (by cycling the active periods of the radio transceiver). The standard aims to facilitate the rapid development of IoT and M2M applications that could exploit burst-like transmissions, such as in metering applications. This type of thinking is very similar to traditional WSN theory and practice, which foreran the development of technologies like 6LoWPAN, RPL, and CoAP, below.

Bluetooth Low Energy (BLE; "Bluetooth Smart") is a recent (2010) integration of Nokia's Wibree (2006) standard with the main Bluetooth standard (originally developed and maintained as IEEE 802.15.1 and Bluetooth SIG). It is designed for short-range (<50 m) applications in healthcare, fitness, security, etc., where high data rates (millions of bits per second) are required to enable application functionality. It is deliberately low cost and energy efficient by design, and has been integrated into the majority of recent smartphones.

Low-Rate, Low-Power Networks are another key technology that form the basis of the IoT. For example, the **IEEE 802.15.4** family of standards was one of the first used in practical research and experimentation in the field of WSNs. It was originally presented in 2003 as Part 15.4: Low-Rate Wireless Personal Area Networks (LR-WPAN). The original release covered the Physical and Medium Access Control layers, specifying use in the ISM bands at frequencies around 433 MHz, 868/915 MHz, and 2.4 GHz. This supported data rates of between 20 kbps up to 256 kbps, depending on selected band, over distances ranging from tens of meters to kilometers (also depending on transmission power level). Typically, radio transceivers developed in compliance with this standard consume power in the tens of milli-Watts range in active modes. This means that they are still insufficiently energy efficient to provide long-lasting battery life (i.e. >10 years) for continuous operation, or energy harvested operation, without aggressively duty cycling the active period of the transceiver. Radio duty cycling refers to managing the active periods of the Radio Frequency Integrated Circuit (RFIC) during transmission, and listening to the medium. It is typically quantified as a percentage with respect to active time.

Notwithstanding, IEEE 802.15.4 defines the PHY layer, and in some instances the MAC layer, upon which a number of low-energy communications specifications have been built. Namely, ZigBee, and its more recent derivatives ZigBee IP and ZigBee RF4CE, WirelessHART, ISA100.a, and others, use this technology at the most basic level.

While the IEEE 802.15.4 standard refers to WPANs in its title, the reality is much different. Recent developments, such as the PHY Amendment for Smart Utility Networks (SUN), IEEE 802.15.4g, seek to extend the operational coverage of these networks up to tens of kilometers in order to provide extremely wide geographic coverage with minimal infrastructure. As the name of the working group suggests, an intuitive use case for this amendment is in the future smart grid.

With **IPv6 Networking**, attention is paid to the ongoing work to facilitate the use of IP to enable interoperability irrespective of the physical and link layers (i.e. making the fact that devices are networked, with or without wires, with various capabilities in terms of range and bandwidth, essentially seamless). It is foreseeable that the only hard requirement for an embedded device will be that it can somehow connect with a compatible gateway device (assuming it is an M2M Area Network device that does not have in-built WAN connectivity, i.e. a cellular modem). The advances in this space have been driven by the IETF from a standardization perspective.

6LoWPAN (IPv6 Over Low Power Wireless Personal Area Networks) was developed initially by the 6LoWPAN Working Group (WG) of the

IETF as a mechanism to transport IPv6 over IEEE 802.15.4-2003 networks. Specifically, methods to handle fragmentation, reassembly, and header compression were the primary objectives. This is due to the large size of IPv6 packets (1280 octets) and the limited space in the protocol data unit (81 octets in the worst case, after security) as a result of the maximum physical packet size of 127 octets, respectively, specified by IEEE 802.15.4-2003. The WG also developed methods to handle address autoconfiguration, the hooks for mesh networking, and network management. We refer the interested reader to the following IETF (open access) requests for comments for detailed information: RFC 4919 (Problem), RFC 6282 (Header Compression), and RFC 6775 (Neighbor Discovery). The WG is currently working on developing IPv6 over Bluetooth Low Energy.

RPL (IPv6 Routing Protocol for Low Power and Lossy Networks) was developed by the IETF Routing over Low Power and Lossy Networks (RoLL) WG. They defined Low Power Lossy Networks as those typically characterized by high data loss rates, low data rates, and general instability. No specific physical or medium access control technologies were specified, but typical links considered include PLC, IEEE 802.15.4, and low-power Wi-Fi. The logic behind the development of the protocol was founded on the traffic flow characteristics of such networks, where typical use cases involve the collection of data from many (for example) sensing points, nodes towards a sink, or alternatively, flooding information from a sink to many nodes in the network. Thus, the well-known concept of a Directed Acyclic Graph (DAG) structure was concentrated to a Destination Oriented DAG (DODAG) for the purposes of initial development. The group defined a new ICMPv6 message, with three possible types, specific for RPL networks. These include a DAG Information Object (DIO), that allows a node to discover an RPL instance, configuration parameters and parents, a DAG Information Solicitation (DIS) to allow requests for DIOs from RPL nodes, and Destination Advertisement Object (DAO), used to propagate destination information upwards (i.e. towards the root) along the DODAG (specific RPL details are available in RFC 6550 and related RFCs). The Trickle Algorithm, standardized in 2011 (RFC 6206), and well known in the WSN community, is an important enabler of RPL message exchange.

CoAP (Constrained Application Protocol) is being developed by the IETF Constrained RESTful Environments (CoRE) WG as a specialized web transfer protocol for use with severe computational and communication constraints typically characteristic of M2M and IoT applications.

Essentially, CoAP elaborates a simple request/response interaction model between application end points (e.g. from an M2M Application to an M2M Area Network Device). REST is essentially a simplification of the ubiquitous HTTP, and thus allows for simple integration between them. This is especially powerful for integrating typical Internet computing applications with constrained devices.

Having covered devices and networking for IoT solutions, we turn our attention to Data Management, which is a core enabling ICT function within IoT systems.

5.3 Data management

5.3.1 Introduction

Modern enterprises need to be agile and dynamically support multiple decision-making processes taken at several levels. In order to achieve this, critical information needs to be available at the right point in a timely manner, and in the right form (Karnouskos 2009). All this info is the result of data being acquired increasingly by M2M interactions, which in conjunction with the processes involved, assist in better decision-making.

Some of the key characteristics of M2M data include:

- **Big Data**: Huge amounts of data are generated, capturing detailed aspects of the processes where devices are involved.
- **Heterogeneous Data**: The data is produced by a huge variety of devices and is itself highly heterogeneous, differing on sampling rate, quality of captured values, etc.
- **Real-World Data**: The overwhelming majority of the M2M data relates to real-world processes and is dependent on the environment they interact with.
- **Real-Time Data**: M2M data is generated in real-time and overwhelmingly can be communicated also in a very timely manner. The latter is of pivotal importance since many times their business value depends on the real-time processing of the info they convey.
- **Temporal Data**: The overwhelming majority of M2M data is of temporal nature, measuring the environment over time.
- **Spatial Data**: Increasingly, the data generated by M2M interactions are not only captured by mobile devices, but also coupled to interactions in specific locations, and their assessment may dynamically vary depending on the location.

- **Polymorphic Data**: The data acquired and used by M2M processes may be complex and involve various data, which can also obtain different meanings depending on the semantics applied and the process they participate in.
- **Proprietary Data**: Up to now, due to monolithic application development, a significant amount of M2M data is stored and captured in proprietary formats. However, increasingly due to the interactions with heterogeneous devices and stakeholders, open approaches for data storage and exchange are used.
- **Security and Privacy Data Aspects**: Due to the detailed capturing of interactions by M2M, analysis of the obtained data has a high risk of leaking private information and usage patterns, as well as compromising security.

In the era of M2M, where billions of devices interact and generate data at exponential growth rates, data management is of critical importance as it sets the basis upon which any other processes can rely and operate. Several aspects of data management need to be addressed in order to fully take advantage of the M2M data and their business relevance.

5.3.2 Managing M2M data

The data flow from the moment it is sensed (e.g. by a wireless sensor node) up to the moment it reaches the backend system has been processed manifold (and often redundantly), either to adjust its representation in order to be easily integrated by the diverse applications, or to compute on it in order to extract and associate it with respective business intelligence (e.g. business process affected, etc.). As depicted also in Figure 5.5, we see a number of data processing network points between the machine and the enterprise that act on the datastream (or simply forwarding it) based on their end-application needs and existing context.

Dealing with M2M data may be decomposed into several stages, which we briefly outline below. Not all of the stages are necessary in every solution, and may be used in orders other than those described below. Additionally, the degree of focus in each stage heavily depends on the actual usage requirements put upon the data as well as the infrastructure.

5.3.2.1 Data generation

Data generation is the first stage within which data is generated actively or passively from the device, system, or as a result of its interactions. The sampling of data generation depends on the device and its capabilities as

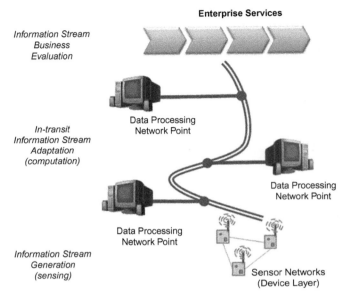

Enterprise Services

Information Stream Business Evaluation

Data Processing Network Point

In-transit Information Stream Adaptation (computation)

Data Processing Network Point

Data Processing Network Point

Sensor Networks (Device Layer)

Information Stream Generation (sensing)

FIGURE 5.5

M2M data from point of generation to business assessment.

well as potentially the application needs. Usually default behaviors for data generation exist, which are usually further configurable to strike a good benefit between involved costs, e.g. frequency of data collection vs. energy used in the case of WSNs, etc. Not all data acquired may actually be communicated as some of them may be assessed locally and subsequently disregarded, while only the result of the assessment may be communicated.

5.3.2.2 Data acquisition

Data acquisition deals with the collection of data (actively or passively) from the device, system, or as a result of its interactions (Karnouskos et al. 2011b). The data acquisition systems usually communicate with distributed devices over wired or wireless links to acquire the needed data, and need to respect security, protocol, and application requirements. The nature of acquisition varies, e.g. it could be continuous monitoring, interval-poll, event-based, etc. The frequency of data acquisition overwhelmingly depends on, or is customized by, the application requirements (or their common denominator).

The data acquired at this stage (for non-closed local control loops) may also differ from the data actually generated. In simple scenarios, due to customized filters deployed at the device, a fraction of the generated data (e.g. adhering to the time of interest or over a threshold) may be communicated. Additionally, in more sophisticated scenarios, data aggregation and even on-device computation of the data may result in communication of key performance indicators of interest to the application, which are calculated based on a device's own intelligence and capabilities (Karnouskos et al. 2011b).

5.3.2.3 Data validation

Data acquired must be checked for correctness and meaningfulness within the specific operating context. The latter is usually done based on rules, semantic annotations, or other logic. Data validation in the era of M2M, where the acquired data may not conform to expectations, is a must as data may be intentionally or unintentionally corrupted during transmission, altered, or not make sense in the business context. As real-world processes depend on valid data to draw business-relevant decisions, this is a key stage, which sometimes does not receive as much attention as it should.

Several known methods are deployed for consistency and data type checking; for example, imposed range limits on the values acquired, logic checks, uniqueness, correct time-stamping, etc. In addition, semantics may play an increasing role here, as the same data may have different meanings in various operating contexts, and via semantics one can benefit while attempting to validate them. Another part of the validation may deal with fallback actions such as requesting the data again if checks fail, or attempts to "repair" partially failed data.

Failure to validate may result in security breaches. Tampered-with data fed to an application is a well known security risk as its effects may lead to attacks on other services, privilege escalation, denial of service, database corruption, etc., as we have witnessed on the Internet over the last decades. As full utilization of this step may require significant computational resources, it may be adequately tackled at the network level (e.g. in the cloud), but may be challenging in direct M2M interactions, e.g. between two resource-constrained machines communicating directly with each other.

5.3.2.4 Data storage

The data generated by M2M interactions is what is commonly referred to as "Big Data." Machines generate an incredible amount of information that is captured and needs to be stored for further processing. As this is proving challenging due to the size of information, a balance between its business usage vs. storage needs to be considered; that is, only the fraction of the data relevant to a business need may be stored for future reference.

This means, for instance, that in a specific scenario, (usually for on-the-fly data that was used to make a decision) once this is done, the processed result can be stored but not necessarily the original data. However, one has to carefully consider what the value of such data is to business not only in current processes, but also potentially other directions that may be followed in the future by the company as different assessments of the same data may provide other, hidden competitive advantages in the future. Due to the massive amounts of M2M data, as well as their envisioned processing (e.g. searching), specialized technologies such as massively parallel processing DBs, distributed file systems, cloud computing platforms, etc. are needed.

5.3.2.5 Data processing

Data processing enables working with the data that is either at rest (already stored) or is in-motion (e.g. stream data). The scope of this processing is to operate on the data at a low level and "enhance" them for future needs. Typical examples include data adjustment during which it might be necessary to normalize data, introduce an estimate for a value that is missing, re-order incoming data by adjusting timestamps, etc. Similarly, aggregation of data or general calculation functions may be operated on two or more data streams and mathematical functions applied on their composition. Another example is the transformation of incoming data; for example, a stream can be converted on the fly (e.g. temperature values are converted from °F to °C), or repackaged in another data model, etc. Missing or invalid data that is needed for the specific time-slot may be forecasted and used until, in a future interaction, the actual data comes into the system. This stage deals mostly with generic operations that can be applied with the aim to enhance them, and takes advantage of low-level (such as DB stored procedures) functions that can operate at massive levels with very low overhead, network traffic, and other limitations.

5.3.2.6 Data remanence

As already discussed, M2M data may reveal critical business aspects, and hence their lifecycle management should include not only the acquisition and usage, but also the end-of-life of data. However, even if the data is erased or removed, residues may still remain in electronic media, and may be easily recovered by third parties — often referred to as data remanence. Several techniques have been developed to deal with this, such as overwriting, degaussing, encryption, and physical destruction. For M2M, points of interest are not only the DBs where the M2M data is collected, but also the points of action, which generate the data, or the individual nodes in between, which may cache it. At the current technology pace, those buffers (e.g. on device) are expected to be less at risk since their limited size means that after a specific time has elapsed, new data will occupy that space; hence, the window of opportunity is rather small. In addition, for large-scale infrastructures the cost of potentially acquiring "deleted" data may be large; hence, their hubs or collection end-points, such as the DBs who have such low cost, may be more at risk. In light of the lack of cross-industry M2M policy-driven data management, it also might be difficult to not only control how the M2M data is used, but also to revoke access to it and "delete" them from the Internet once shared.

5.3.2.7 Data analysis

Data available in the repositories can be subjected to analysis with the aim to obtain the information they encapsulate and use it for supporting decision-making processes. The analysis of data at this stage heavily depends on the domain and the context of the data. For instance, business intelligence tools process the data with a focus on the aggregation and key performance indicator assessment. Data mining focuses on discovering knowledge, usually in conjunction with predictive goals. Statistics can also be used on the data to assess them quantitatively (descriptive statistics), find their main characteristics (exploratory data analysis), confirm a specific hypothesis (confirmatory data analysis), discover knowledge (data mining), and for machine learning, etc. This stage is the basis for any sophisticated applications that take advantage of the information hidden directly or indirectly on the data, and can be used, for example, for business insights, etc. M2M has the potential to revolutionize modern businesses, and we analyze some of these aspects more in Sections 5.6 and 5.7, where we focus on data science and knowledge management.

5.3.3 Considerations for M2M data

We have already provided insights on the M2M data management from cradle to grave. However, the real paradigm change the M2M data will enforce depends on a single aspect: data sharing. Although there are benefits acquired by processing the M2M data at local loops, their real benefit is brought into the foreground when these are shared at large scale. The latter can act as an enabler to better understand complex systems of systems, and better manage them. The Cooperating Objects vision (Marrón et al. 2011), which assumes cooperation among devices and systems as the key driving force for interaction, sheds some light on the benefits and challenges that will emerge in all layers of such an M2M infrastructure.

As an indicative example, the smart city can be used where huge data from its infrastructure, citizens, businesses, and individual assets need to be considered, analyzed, and after decisions are taken, enforced. M2M data hold the key to do so, and enable an efficient and sustainable future.

The M2M infrastructure in place heavily depends on real-world processes, implying also that a big percentage of data will be generated by machines that interact with the real-world environment, while the rest will be purely virtual data. For the first part, where machines are involved, there is a real cost for the infrastructure that has to be met. Hence, it is expected that stakeholders in the future will further diversify, and we will see the emergence of infrastructure providers who will operate and manage many of the machines generating this data, which can then be communicated to others (e.g. analytics specialists to take advantage of the insights offered). The end-beneficiaries might acquire information, but do not necessarily need to have access or to process the data by themselves. Hence, as we see, there is a rise of specialists in the various stages of M2M data management that will cooperate with application providers, users, etc. for the common benefit. Such ecosystems are expected to be of key importance in the future IoT era. This transition is already at an early stage, and boldly contradicts the existing initial M2M efforts, where the application developer, the data collector, and the infrastructure operator roles are largely performed by the same stakeholder (or a very small number of them).

Because of expected wide sharing of data and usage in multiple applications, security and trust are of key importance. Security is mandatory for enabling confidentiality, integrity, availability, authenticity, and non-repudiation of data from the moment of generation to consumption. Due to the large-scale IoT infrastructure, heterogeneous devices, and stakeholders involved, this will be challenging. In addition, trust will be another major issue, as even if data is securely communicated or verified, the level of trust based on them will impact the decision-making process and risk analysis. The recent replaying of data in the Stuxnet worm (Karnouskos 2011a) has demonstrated that although data may appear legitimate, they still need to be independently verified to make sure that the whole chain from generation to consumption is not tampered with. Managing security and trust in the highly federated M2M-envisioned infrastructures poses a significant challenge, especially for mission critical applications that also exercise control.

Privacy is also expected to be a significant issue in IoT infrastructures. Currently, a lot of emphasis is put on acquiring the data, and no real solutions exist for large-scale systems to share data in a controlled way. Once data is shared, the originator has no more control over its lifetime. This calls for policy-driven data management for their whole lifecycle so that, for instance, data can be invalidated, or even removed from the IoT global ecosystem once wished so, as we have indicated in the data remanence stage. A typical example here constitutes the usage of private citizen data, which could be controllably shared as wished; it should also be possible to (partially) revoke that right at will. Understandably, this is an issue that will not be solved only by technology, but will need to be accompanied by the appropriate legislation frameworks.

Finally, we are still at the dawn of an era that has to deal with huge data and unveil the hidden information patterns behind them. Being able to search and apply intelligent algorithms that may unveil those hidden patterns is expected to be a significant business advantage. Data Science in the IoT era is a cross-discipline approach building on mathematics, statistics, high-performance computing, modeling, machine learning, engineering, etc. that will play a key role in understanding the data, assessing their information at large scale, and hopefully enabling the better studying of complex systems of systems and their emergent characteristics. The latter brings hope that features of several real-world processes in the context of IoT, such as cascading failures, dynamic behaviors, non-linear relationships, feedback-loops, nested systems, etc., will be better studied and understood and applied in real-world domains (e.g. smart city, markets, enterprises, planet ecosystem, etc.).

5.3.4 Conclusions

Data and its management hold the key to unveiling the true power of M2M and IoT. To do so, however, we have to think and develop approaches that go beyond simple data collection, and enable the management of their whole lifecycle at very large scale, while in parallel considering the special needs and the usage requirements posed by specific domains or applications. Mastering the challenges of data management will enable data analysis to flourish, and this in turn will empower new innovative approaches to be realized for the benefit of citizens, business, and society.

5.4 Business processes in IoT
5.4.1 Introduction

A business process refers to a series of activities, often a collection of interrelated processes in a logical sequence, within an enterprise, leading

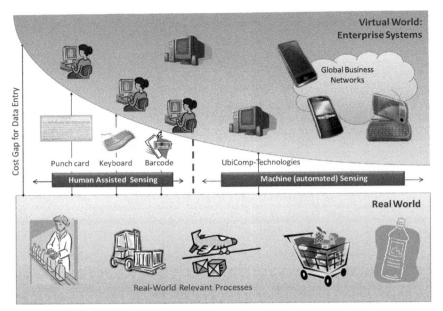

FIGURE 5.6

The decreasing cost of information exchange between the real-world and enterprise systems with the advancement of M2M.

to a specific result. There are several types of business processes such as management, operational, and supporting, all of which aim at achieving a specific mission objective. As business processes usually span several systems and may get very complex, several methods and techniques have been developed for their modeling, such as the Business Process Model and Notation (BPMN), which graphically represents business processes in a business process model. Managers and business analysis model an enterprise's processes in an effort to depict the real way an enterprise operates and subsequently to improve efficiency and quality.

Several key business processes in modern enterprise systems heavily rely on interaction with real-world processes, largely for monitoring, but also for some control (management), in order to take business-critical decisions and optimize actions across the enterprise. The introduction of modern ICT has significantly changed the way enterprises (and therefore business processes) interact with the real world.

As depicted in Figure 5.6, we have witnessed a paradigm change with the dramatic reduction of the data acquisition from the real world; this was attributed mostly to the automation offered by machines embedded in the real world. Initially all these interactions were human-based (e.g. via a keyboard) or human-assisted (e.g. via a barcode scanner); however, with the prevalence of RFID, WSNs, and advanced networked embedded devices, all information exchange between the real-world and enterprise systems can be done automatically without any human intervention and at blazing speeds.

In the M2M era, connected devices can be clearly identified, and with the help of services, this integration leads to active participation of the devices to the business processes. This direct integration is changing the way business processes are modeled and executed today as new requirements come into play. Existing modeling tools are hardly designed to specify aspects of the real world in modeling environments and capture their full characteristics. To this direction, the existence of SOA-ready devices (i.e. devices that offer their functionalities as a web service) simplifies the integration and interaction as they can be considered as a traditional web service that runs on a specific device. Nevertheless, there is ongoing research on inclusion of semantics in order to include, in an easier and more accurate way, the M2M in business process modeling and execution.

The industrial adoption of IoT (e.g. of wireless sensor networks) is hampered by the lack of integration of WSNs with business process modeling languages and back-end systems.

There are, however, promising approaches such as the one provided by makeSense (Tranquillini et al. 2012), which tackles this problem space with a unified programming framework and a compilation chain that, from high-level business process specifications, generates code ready for deployment on WSN nodes. A layered approach for developing, deploying, and managing WSN applications that natively interact with enterprise information systems such as a business process engine and the processes running therein is proposed and assessed.

M2M and IoT empower business processes to acquire very detailed data about the operations, and be informed about the conditions in the real world in a very timely manner. Subsequently, better business intelligence (Spiess & Karnouskos 2007) and more informed decision-making can be realized. The latter enables businesses to operate more efficiently, which translates to a business competitive advantage.

5.4.2 IoT integration with enterprise systems

M2M communication and the vision of the IoT pose a new era where billions of devices will need to interact with each other and exchange information in order to fulfill their purpose. Much of this communication

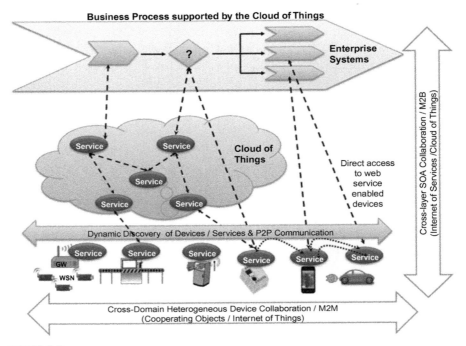

FIGURE 5.7

A collaborative infrastructure driven by M2M and M2B.

is expected to happen over Internet technologies (Vasseur and Dunkels 2010) and tap into the extensive experience acquired with architectures and experiences in the Internet/Web over the last several decades. More sophisticated, though still overwhelmingly experimental, approaches go beyond simple integration and target more complex interactions where collaboration of devices and systems is taking place.

As shown in Figure 5.7, cross-layer interaction and cooperation can be pursued:

- at the M2M level, where the machines cooperate with each other (machine-focused interactions), as well as
- at the machine-to-business (M2B) layer, where machines cooperate also with network-based services, business systems (business service focus), and applications.

As depicted in Figure 5.7, we can see several devices in the lowest layer. These can communicate with each other over short-range protocols (e.g. over ZigBee, Bluetooth), or even longer distances (e.g. over Wi-Fi, etc.).

Some of them may host services (e.g. REST services), and even have dynamic discovery capabilities based on the communication protocol or other capabilities (e.g. WS-Eventing in DPWS). Some of them may be very resource constrained, which means that auxiliary gateways could provide additional support such as mediation of communication, protocol translation, etc. Independent of whether the devices are able to discover and interact with other devices and systems directly or via the support of the infrastructure, the M2M interactions enable them to empower several applications and interact with each other in order to fulfill their goals.

Promising real-world integration is done using a service-oriented approach by interacting directly with the respective physical elements, for example, via web services running on devices (if supported) or via more lightweight approaches such as REST. In the case of legacy systems, gateways and service mediators are in place to enable such integration challenges. However, in the era of sophisticated networked embedded devices, open interactions, and virtualized resources, a dilemma is emerging which manifests to which functionalities can be abstracted and migrated to the cloud (with all the benefits and constraints it poses), and which of these should still remain on the device itself. This is not an easy decision to take, as benefits and constraints exist in both directions, so practically, we expect an amalgamation of them to occur (Karnouskos & Somlev 2013).

Many of the services that will interact with the devices are expected to be network services available, for example, in the cloud. The main motivation for enterprise services is to take advantage of the cloud characteristics such as virtualization, scalability, multi-tenancy, performance, lifecycle management, etc. Similarly, we expect to see that a large number of devices, and generally cyber-physical systems, will make their functionality available on the cloud (Karnouskos & Somlev 2013). As such, we are moving towards an infrastructure where the cloud and its services (as depicted in Figure 5.8) take a prominent position towards empowering modern enterprises and their business processes.

A key motivator is the minimization of communication overhead with multiple endpoints by, for example, transmission of data to a single or limited number of points in the network, and letting the cloud do the load-balancing and further mediation of communication. For instance, as depicted in the figure above, a Content Delivery Network (CDN) can be used in order to get access to the generated data from locations that are far away from the M2M infrastructure (geographically, network-wise, etc.).

To this end, the data acquired by the device can be offered without over-consumption of the device's resources, while in parallel, better control and

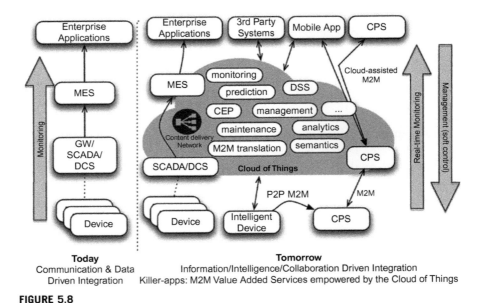

FIGURE 5.8

The Cloud of Things as an enabler for new value added services & apps.

management can be applied. Typical examples include enabling access to the full historical data, preprocessing of information, transparently upgrading the cloud services, or even not providing access to internal systems for security reasons. This clear decoupling of "things" and the usage of their data is expected to further empower information-driven business processes and applications that can operate over federated infrastructures.

5.4.3 Distributed business processes in IoT

Today, as seen on the left part of Figure 5.9, the integration of devices in business processes merely implies the acquisition of data from the device layer, its transportation to the backend systems, its assessment, and once a decision is made, potentially the control (management) of the device, which adjusts its behavior. However, in the future, due to the large scale of IoT, as well as the huge data that it will generate, such approaches are not viable.

Transportation of data from the "point of action" where the device collects or generates them, all the way to the backend system to then evaluate their usefulness, will not be practical for communication reasons, as well as due to the processing load that it will incur at the enterprise side; this is

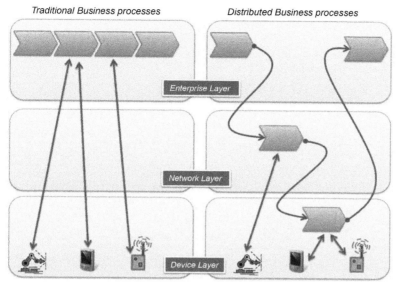

FIGURE 5.9

Distributed Business Processes in M2M era.

something that the current systems were not designed for. Enterprise systems trying to process such a high rate of non- or minor-relevancy data will be overloaded.

As such, the first strategic step is to minimize communication with enterprise systems to only what is relevant for business. With the increase in resources (e.g. computational capabilities) in the network, and especially on the devices themselves (more memory, multi-core CPUs, etc.), it makes sense not to host the intelligence and the computation required for it only on the enterprise side, but actually distribute it on the network, and even on the edge nodes (i.e. the devices themselves), as depicted on the right side of Figure 5.9.

Partially outsourcing functionality traditionally residing in backend systems to the network itself and the edge nodes means we can realize distributed business processes whose sub-processes may execute outside the enterprise system. As devices are capable of computing, they can either realize the task of processing and evaluating business relevant information they generate by themselves or in clusters.

Distributing the computational load in the layers between enterprises and the real-world infrastructure is not the only reason; distributing business intelligence is also a significant motivation. Business processes can

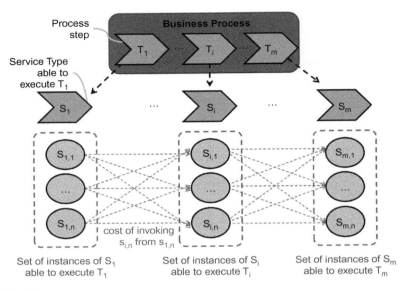

FIGURE 5.10

On-Device and in-network Business Process Composition and runtime execution.

bind during execution of dynamic resources that they discover locally, and integrate them to better achieve their goals. Being in the world of service mash-ups, we will witness a paradigm change not only in the way individual devices, but also how clusters of them, interact with each other and with enterprise systems.

Modeling of business processes (Spiess et al. 2009) can now be done by focusing on the functionality provided and that can be discovered dynamically during runtime, and not on the concrete implementation of it; we care about what is provided but not how, as depicted in Figure 5.10. As such, we can now model distributed business processes that execute on enterprise systems, in-network, and on-device. The vision (Spiess et al. 2009) is to additionally consider during runtime the requirements and costs associated with the execution in order to select the best of available instances and optimize the business process in total according to the enterprise needs, e.g. for low impact on a device's energy source, or for high-speed communication, etc.

5.4.4 Considerations

Existing tools and approaches need to be extended to the make the business processes IoT aware. The current terminology in modeling tools is focused on the enterprise context and does not include the notation of physical entities such as devices as these are considered in IoT. Although distributed execution of processes exists (e.g. in BPMN), additional work is needed to be able to select the devices in which such processes execute and consider their characteristics or dynamic resources, etc. The dynamic aspect is of key importance in the IoT, as this is mobile and availability is not guaranteed, which means that availability in modeling time does not guarantee availability at runtime and vice-versa. Even if the latter holds true, this might again change during the execution of a business process; hence, fault tolerance needs to be considered.

IoT infrastructures are expected to be of large scale. Hence, scalability is an aspect that needs to be considered in the business process modeling and execution. In addition, event-based interactions among the processes play a key role in IoT, as a business process flow may be influenced by an event, or as its result, trigger a new event. Such considerations need to be seen also under the light of real-time interactions, which may be mandatory in several domains (e.g. industrial automation). The same also holds true for the quality of information acquired at each step, as resource-constrained devices may have a higher probability of delivering non-error-free information (e.g. due to malfunctions, etc.). The latter would enable a different modeling of processes that depend on this information, and not consider it always as correct or trustworthy. To solve this today, additional data management steps are considered, which we described in Section 5.3.

5.4.5 Conclusions

Modern enterprises operate on a global scale and depend on complex business processes. Business continuity needs to be guaranteed, and therefore efficient information acquisition, evaluation, and interaction with the real world are of key importance. The infrastructure envisioned is a heterogeneous one, where millions of devices are interconnected, ready to receive instructions and create event notifications, and where the most advanced ones depict self-behavior (e.g. self-management, self-healing, self-optimization, etc.) and collaborate. This can lead to a paradigm change as business logic can now be intelligently distributed to several layers such as the network, or even the device layer, creating new opportunities, but also challenges that need to be assessed. Future Enterprise systems will be in position to better integrate state and events of the physical world in a timely manner, and hence to lead to more diverse, highly dynamic, and efficient business applications.

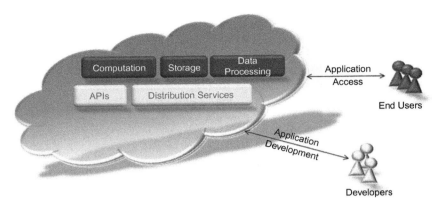

FIGURE 5.11

Conceptual Overview of Cloud Computing.

5.5 Everything as a service (XaaS)

As discussed in Chapter 1, there is a general trend away from locally managing dedicated hardware toward cloud infrastructures that drives down the overall cost for computational capacity and storage. This is commonly referred to as "cloud computing." Cloud computing is a model for enabling ubiquitous, on-demand network access to a shared pool of configurable computing resources (e.g. networks, servers, storage, applications, and services) that can be provisioned, configured, and made available with minimal management effort or service provider interaction.

Cloud computing, however, does not change the fundamentals of software engineering. All applications need access to three things: compute, storage, and data processing capacities. With cloud computing, a fourth element is added — distribution services — i.e. the manner in which the data and computational capacity are linked together and coordinated.

A cloud-computing platform may therefore be viewed conceptually (Figure 5.11).

Several essential characteristics of cloud computing have been defined by NIST (2011) as follows:

- *On-Demand Self-Service.* A consumer can unilaterally provision computing capabilities, such as server time and network storage, as needed, or automatically, without requiring human interaction with each service provider.

- *Broad Network Access.* Capabilities are available over the network and accessed through standard mechanisms that promote use by heterogeneous thin or thick client platforms (e.g. mobile phones, tablets, laptops, and workstations).
- *Resource Pooling.* The provider's computing resources are pooled to serve multiple consumers using a multi-tenant model, with different physical and virtual resources dynamically assigned and reassigned according to consumer demand. There is a sense of location independence in that the customer generally has no control or knowledge over the exact location of the provided resources, but may be able to specify location at a higher level of abstraction (e.g. country, state, or datacenter). Examples of resources include storage, processing, memory, and network bandwidth.
- *Rapid Elasticity.* Capabilities can be elastically provisioned and released, in some cases automatically, to scale rapidly outward and inward commensurate with demand. To the consumer, the capabilities available for provisioning often appear to be unlimited, and can be appropriated in any quantity at any time.
- *Measured Service.* Cloud systems automatically control and optimize resource use by leveraging a metering capability, at some level of abstraction, appropriate to the type of service (e.g. storage, processing, bandwidth, and active user accounts). Resource usage can be monitored, controlled, and reported, providing transparency for both the provider and consumer of the utilized service.

Within the mobile industry, there is an increasing interest in this space; for example, Network Function Virtualization (NFV) allows even mobile operators to move features into cloud-based infrastructures. It is important to note that cloud computing is currently an evolving paradigm (NIST 2011).

Once such infrastructures are available, however, it is easier to deploy applications in software. For M2M and IoT, these infrastructures provide the following:

1. Storage of the massive amounts of data that sensors, tags, and other "things" will produce.
2. Computational capacity in order to analyze data rapidly and cheaply.
3. Over time, cloud infrastructure will allow enterprises and developers to share datasets, allowing for rapid creation of information value chains as described in Chapter 1.

Cloud computing comes in several different service models and deployment options for enterprises wishing to use it. The three main service models may be defined as (NIST 2011):

Software as a Service (SaaS): Refers to software that is provided to consumers on demand, typically via a thin client. The end-users do not manage the cloud infrastructure in any way. This is handled by an Application Service Provider (ASP) or Independent Software Vendor (ISV). Examples include office and messaging software, email, or CRM tools housed in the cloud. The end-user has limited ability to change anything beyond user-specific application configuration settings.

Platform as a Service (PaaS): Refers to cloud solutions that provide both a computing platform and a solution stack as a service via the Internet. The customers themselves develop the necessary software using tools provided by the provider, who also provides the networks, the storage, and the other distribution services required. Again, the provider manages the underlying cloud infrastructure, while the customer has control over the deployed applications and possible settings for the application-hosting environment (NIST 2011).

Infrastructure as a Service (IaaS): In this model, the provider offers virtual machines and other resources such as hypervisors (e.g. Xen, KVM) to customers. Pools of hypervisors support the virtual machines and allow users to scale resource usage up and down in accordance with their computational requirements. Users install an OS image and application software on the cloud infrastructure. The provider manages the underlying cloud infrastructure, while the customer has control over OS, storage, deployed applications, and possibly some networking components.

Deployment Models:

Private Cloud. The cloud infrastructure is provisioned for exclusive use by a single organization comprising multiple consumers (e.g. business units). It may be owned, managed, and operated by the organization, a third party, or some combination of them, and it may exist on or off premises.

Community Cloud. The cloud infrastructure is provisioned for exclusive use by a specific community of consumers from organizations that have shared concerns (e.g. mission, security requirements, policy, and compliance considerations). It may be owned, managed, and operated by one or more of the organizations in the community, a third party, or some combination of them, and it may exist on or off premises.

Public Cloud. The cloud infrastructure is provisioned for open use by the general public. It may be owned, managed, and operated by a business, academic, or government organization, or some combination thereof. It exists on the premises of the cloud provider. *Hybrid Cloud.* The cloud infrastructure is a composition of two or more distinct cloud infrastructures (private, community, or public) that remain unique entities, but are bound together by standardized or proprietary technology that enables data and application portability (e.g. cloud bursting for load balancing between clouds).

We now turn our attention to analytics for M2M and IoT solutions.

5.6 M2M and IoT analytics

5.6.1 Introduction

Traditionally, M2M data has been sent from specific devices to specific services, which store the data of interest. This approach uses semantically well-defined data for specific purposes, and only requires storing the data that is needed for the explicit use cases, and only for as long as it's required. For the most part, the applications have been monitoring, reporting, and rule-based actions.

To further increase the speed of M2M deployments, it's important to look at methods to extract additional value from these devices. Given the enormous amounts of data that will be generated by the IoT and the advancements within the area of Big Data, new opportunities arise from the possibility to reuse data from devices for multiple purposes, many of which will not even be imagined at the time of deployment. The opportunities of using M2M data for advanced analytics and business intelligence are very promising. By transforming raw data into actionable intelligence, it's possible to improve many areas, such as enhancement of existing products, cost-savings, service quality, as well as operational efficiency.

By applying technologies from the Big Data domain, it is possible to store more data, such as contextual and situational information, and given a more open approach to data, such as the open-data government initiatives (e.g. Data.gov and Data.gov.uk), even more understanding can be derived, which can be used to improve everything from Demand/Response in a power grid to wastewater treatment in a city. This requires a migration from the data silos of today to an architecture where it's possible to cross-analyze data residing in many different locations, which implies that the location of the stored data will not necessarily be the same as the location where the analytics will take place. From a practical standpoint, this is a problem that needs to be handled when we are talking about extreme amounts of data.

Descriptive statistics can take you a long way from raw data to actionable intelligence. Other opportunities are provided by data mining and machine learning, with no clear distinction between the three, although data mining can be described as the automatic or semiautomatic task of extracting previously unknown information from a large quantity of data, while machine learning is focused on finding models for specific tasks, e.g. separate spam from non-spam email.

For M2M data, traditional data warehousing and analytics will for many cases not be up to the task. Big Data technologies such as MapReduce for massively parallel analytics, as well as analytics on online streaming data where the individual data item is not necessarily stored, will play an important role in the management and analysis of large-scale M2M data.

To handle the analytical needs related to M2M and IoT, it's expected that in the near term, vendors of Big Data solutions will provide for the needs of in-house analytics. In the long term, new niches are likely to appear, such as cloud storage providers, data brokers, and Analytics-as-a-Service providers. Apart from the software and services provided for analytics, a major uptake in professional services for consultancy within M2M analytics is expected (Figure 5.12).

The revenues within M2M analytics are expected to grow rapidly for most industry verticals (Figure 5.13).

In this chapter we will cover some general purposes of M2M analytics, the process involved in analytics, as well as provide an architectural overview.

5.6.2 Purposes and considerations

Regardless of whether you call it statistics, data mining, or machine learning, there exist a multitude of methods to extract different types of information from data. The information can be used in everything from static reports to interactive decision support systems, or even fully automated real-time systems.

Some examples of methods and purposes are as follows:

- **Descriptive Analytics**: Use of means, variances, maxima, minima, aggregates, and frequencies, optionally grouped by selected characteristics.
  - Create Key Performance Indicators (KPI's) that enable better understanding of the performance of complex systems such as cellular networks or oil pipelines.

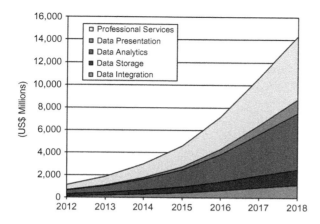

FIGURE 5.12

M2M Analytics Revenues by Segment, World Market, Forecast: 2012 to 2018.

(ABI Research 2013)

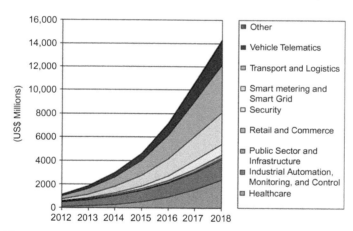

FIGURE 5.13

M2M Analytics Revenues by Industry, Vertical World Market, Forecast: 2012 to 2018.

(ABI Research 2013)

- **Predictive Analytics**: Use current and historical facts to predict what will happen next.
  - Forecast demand and supply in a power grid and train a model to predict how price affects electric usage to optimize the performance and minimize peaks in the electricity consumption.
  - Predictive maintenance on electromechanical equipment in a nuclear power plant by modeling the relationship between device health characteristics measured by sensors and historic failures.

- Understand how electricity and water consumption relates to regional demographics.
- Model the effects of traffic lights on a city's road network based on data from cars and sensors in the city to minimize congestion.
- **Clustering**: Identification of groups with similar characteristics.
  - Perform customer segmentation or find behavioral patterns in a large set of M2M devices.
  - Mine time series data for recurring patterns that can be used in predictive analytics to detect, for example, fraud, machine failures, or traffic accidents.
- **Anomaly Detection**:
  - Detect fraud for smart meters by checking for anomalous electricity consumption compared to similar customers, or historic consumption for the subscriber.

There are some considerations in regards to analyzing M2M data. M2M analytical solutions can use data from any kind of source, e.g. dynamic data such as individual sensors, mobile device sensors, and social networks, as well as more static data that are relevant for analysis, e.g. Geographic Information System (GIS) data, public city data, or national statistics. These can be divided into two categories, enterprise specific data and public data, of which the former can be efficiently accessed using common formatting, processing, and storage. The latter is most often accessed using public APIs, and has commonalities to a much lesser degree.

M2M data fulfills all the characteristics of Big Data, which is usually described by the four "Vs":

- **Volume**: To be able to create good analytical models it's no longer enough to analyze the data once and then discard it. Creating a valid model often requires a longer period of historic data. This means that the amount of historic data for M2M devices is expected to grow rapidly.
- **Velocity**: Even though M2M devices usually report quite seldom, the sheer number of devices means that the systems will have to handle a huge number of transactions per second. Also, often the value of M2M data is strongly related to how fresh it is to be able provide the best actionable intelligence, which puts requirements on the analytical platform.
- **Variation**: Given the multitude of device types used in M2M, it's apparent that the variation will be very high. This is further complicated by the use of different data formats as well as different

configurations for devices of the same type (e.g. where one device measures temperature in Celsius every minute, another device measures it in Fahrenheit every hour). The upside is that the data is expected to be semantically well-defined, which allows for simple transformation rules.

- **Veracity**: It's imperative that we can trust the data that is analyzed. There are many pitfalls along the way, such as erroneous timestamps, non-adherence to standards, proprietary formats with missing semantics, wrongly calibrated sensors, as well as missing data. This requires rules that can handle these cases, as well as fault-tolerant algorithms that, for example, can detect outliers (anomalies).

Last but not least are the consequences of the need for user privacy. This means that data will often be anonymized both in terms of removing user identities, as well as uniqueness of user data. This limits the possibilities of cross-referencing different data sources.

5.6.3 **Analytics architecture**

An architecture for analytics needs to take a few basic requirements into account (Figure 5.14). One of these is to serve as a platform for data exploration and modeling by data scientists and other advanced information consumers performing business analytics and intelligence. As much time is spent on data preparation before any analytics can take place, this is also an integral part of the architecture to facilitate. Finally, efficient means of building and viewing reports, as well as integrating with back-end systems and business processes, is of importance. These requirements concern batch analytics, but should also be considered for stream analytics.

Note that an analytics architecture is not intended for general-purpose data storage, although sometimes it's efficient to co-locate these two functions into one architecture. Risks of affecting production must, however, be taken into consideration if this is done instead of importing the data into an analytics sandbox where analysts can work on the data independently. Another benefit with an analytics sandbox is also that this environment offers a full suite of analytical tools that normally cannot be found in a traditional database. It also offers a development platform with the necessary computing resources required to perform complex analytics on very big data sets.

A sandbox for Big Data analytics can be realized in a number of ways, of which the Hadoop ecosystem is probably the best known.

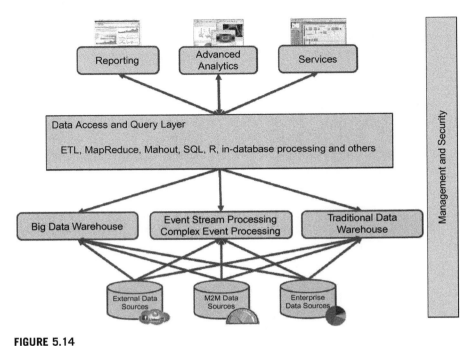

FIGURE 5.14

Analytics Architectural Overview.

Other alternatives include:

- Columnar databases such as HP Vertica, Actian ParAccel MPP, SAP Sybase IQ, and Infobright.
- Massively Parallel Processing (MPP) architectures such as Pivotal Greenplum and Teradata Aster.
- In-memory databases such as SAP Hana and QlikView.

All of the above focus on batch-oriented analytics, where all data is available for the model generation. A complimentary method is to perform analytics on the live data streams (i.e. stream analytics), which means that the data does not need to be stored after it has been processed. This in turn limits the available algorithms to those that can handle incremental model building. The most common technologies in this segment are Event Stream Processing (e.g. Twitter Storm and Apache S4) and Complex Event Processing (e.g. EsperTech Esper and SAP Sybase Event Stream Processor).

An analytical architecture should preferably also provide:

- Authentication and authorization to access data.
- Failover and redundancy features.

- Management facilities.
- Efficient batch loading of data and support self-service.
- Scheduling of batch jobs, such as data import and model training.
- Connectors to import data from external sources.

Since the Hadoop ecosystem is so prevalent in the area of Big Data and analytics, it's worth going into a bit more depth. The core of Hadoop is the MapReduce programming model, which allows processing of large data sets by deploying an algorithm, written as a program, onto a cluster of nodes. A MapReduce job reads data from the Hadoop File System (HDFS), and runs on the same nodes as the deployed algorithm. This allows the Hadoop framework to utilize data locality as much as possible to avoid unnecessary transfer of data between the nodes. MapReduce is batch-oriented and intended for very large jobs that typically take an hour or more to execute. The nodes and services in a Hadoop cluster are coordinated by ZooKeeper, which serves as a central naming and configuration service.

Although it's not unusual for developers to use MapReduce directly, there exist a number of technologies that provide further abstraction levels, such as:

- **HBase**: A column-oriented data store that provides real-time read/write access to very large tables distributed over HDFS.
- **Mahout**: A distributed and scalable library of machine learning algorithms that can make use of MapReduce.
- **Pig**: A tool for converting relational algebra scripts into MapReduce jobs that can read data from HDFS and HBase.
- **Hive**: Similar to Pig, but offers an SQL-like scripting language called HiveQL instead.
- **Impala**: Offers low-latency queries using HiveQL for interactive exploratory analytics, as compared to Hive, which is better suited for long running batch-oriented tasks.

It is worth noting that both Hadoop and the other technologies mentioned in this chapter are evolving rapidly and that current limitations are quickly being addressed by adding new features and by borrowing technologies from each other.

5.6.4 Methodology

Knowledge discovery and analytics can be described as a project methodology, following certain steps in a process model. To perform efficient

Table 5.2 Summary of the Correspondences between KDD, SEMMA, and CRISP-DM

KDD	SEMMA	CRISP-DM
Pre-KDD	—	Business understanding
Selection	Sample	Data understanding
Pre-processing	Explore	
Transformation	Modify	Data preparation
Data mining	Model	Modeling
Interpretation/Evaluation	Assessment	Evaluation
Post-KDD	—	Deployment

Source: Azevedo & Santos (2008).

analytics and find answers to important questions, it's paramount to involve the right people with the necessary business understanding at the beginning of a project. When the goals have been understood, the next step is to gather the necessary data and to understand it in terms of characteristics and quality. When this is done, it's possible to build the models that can answer the previously stated questions, although quite commonly the data needs to be transformed first. Before a model is deployed in the organization, it's also important to evaluate the performance of the model, i.e. How well does it fare in the real world?

A model that has been developed can be used in a multitude of scenarios, such as an executive performing Enterprise Performance Management, information consumers using fixed reports, or an automated process deciding on individual transactions using a knowledge-driven business process.

There exist several process models that include some or all parts of the steps mentioned above, such as the *Knowledge Discovery in Databases* (KDD) process, or the industrial standards *Sample, Explore, Modify, Model and Assess* (SEMMA), and *Cross Industry Standard Process for Data Mining* (CRISP-DM) (Table 5.2). The most commonly used process model of these is CRISP-DM (Kdnuggets 2007).

The phases in the CRISP-DM process model are described in Figure 5.15, which is followed by descriptions of each of the phases. These are illustrated using an example from Predictive Maintenance (PdM) for pump stations in a water distribution network. Although the figure indicates a certain order between the phases, analytics is an iterative process, and it's expected that you will have to move back and forth between the phases to a certain extent.

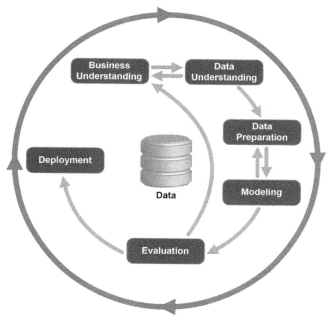

FIGURE 5.15

CRISP-DM Process Diagram.

(Wikimedia Commons 2013)

5.6.4.1 Business understanding

The first phase in the process is to understand the business objectives and requirements, as well as success criteria. This forms the basis for formulating the goals and plan for the data mining process.

Many organizations may have a feeling that they are sitting on valuable data, but are unsure how to capitalize on this. In these cases, it's not unusual to bring in the help of an analytics team to identify potential business cases that can benefit from the data.

5.6.4.1.1 Predictive maintenance example

It has been decided to start a project with the main objective to study ways to reduce the frequency of costly unplanned emergency repair work and downtime in pumping stations by predicting future pump failures or necessary maintenance work in good time to allow for organized maintenance to take place.

The project will be determined a success if the study evaluation shows that:

- Pump station downtime is reduced by 10%.
- Maintenance costs for pump stations is reduced by 15%.
- The project is concluded within time and kept on budget.

These business requirements are translated into more practical data mining goals, such as evaluating two different approaches to predicting maintenance actions:

- **Action Forecasting**: Train a model that can predict needed maintenance actions based on vibrations and other pump characteristics. Apply forecasting methods on the vibration measurement sensors to predict future maintenance actions.
- **Similar Case Recommendations**: Use information about pumps, such as manufacturer, model, and age, as well as information about working conditions, such as workload, water corrosiveness, and percentage of sand and grit, to define groups of similar pumps. Use data from prior pump failures of similar pumps, as well as prior maintenance decisions, to recommend actions to take.

5.6.4.2 Data understanding
The next phase consists of collecting data and gaining an understanding of the data properties, such as amount of data and quality in terms of inconsistencies, missing data, and measurement errors. The tasks in this phase also include gaining some understanding of actionable insights contained in the data, as well as to form some basic hypotheses.

5.6.4.2.1 Predictive maintenance example
In this example, there are several data sources that could be of interest, including:

- Trouble Reports from the Ticket System: These will allow us to correlate prior failures with historic data from the pumps.
- Work Orders: Information from these enable us to understand what actions were taken in regards to failing/failed pumps.
- Pump Information: Descriptive data such as water source, manufacturer, model, and age of pump and it's major parts, as well as information about working conditions (e.g. water corrosiveness, percentage of sand and grit).
- Sensor Data: Historic measurements of pump revolutions per minute (RPM) and vibration measurements.

The different data sets are then explored and described accordingly:

- An inspection of the data shows that there are several million records of sensor data, spanning from approximately a two-year time period.
- Most of the data, apart from the sensor data, is symbolic, e.g. manufacturer and equipment identifiers. Some fields are numerical, e.g. percentage of sand, equipment age, and vibrations. There is also some unstructured data in the form of free text descriptions in the trouble reports and work orders.
- The pump information sometimes contains a pump identifier in the form of a numeric key, and sometimes using a textual name. The same problem is observed for the water source identifier. These are key fields that are used for mapping the different data sources with each other, and some work to fix them will be needed.
- In some cases there are missing data, both from sensor data (e.g. when a station has been services) and from the contextual data, e.g. missing pump information.
- It is noted that the measurements from the sensors indicates that the sensors in some cases have not been calibrated correctly.

5.6.4.3 Data preparation

Before it's possible to start modeling the data to achieve our goals, it's necessary to prepare the data in terms of selection, transformation, and cleaning. In this phase, it's frequently the case that new data is necessary to construct, both in terms of entirely new attributes as well as imputing new data into records where data is missing.

It's quite common for this phase to consume more than half the time of a project.

5.6.4.3.1 Predictive maintenance example

Several operations are performed to prepare the data, and three data sets are constructed:

- Vibration Time Series
  - Time series data for the pump vibrations are at the core of the analytical work. Missing values are estimated and imputed to create complete time series representations. The measurement values are adjusted to account for incorrectly calibrated sensors.
- Workload Time Series
  - Pump RPM measurements are used to construct a new time series with attributes that describe the pump workload at a given date,

  e.g. average daily workload, standard deviation of the workload, maximal daily workload, and workload trend.
- Pump Records
  - Information needed for grouping similar pumps is included and joined with the newly created workload data.
- Action Records
  - Trouble reports and work orders are joined to create one action record for each maintenance task. Some of the data is excluded and transformed to create new attributes that indicate what kind of action was performed, e.g. bearing replacement, oil lubrication, or motor replacement.
  - The action records are merged with the pump records, as they were at the time the action was performed.

5.6.4.4 Modeling

At the modeling phase, it's finally time to use the data to gain an under-standing of the actual business problems that were stated in the beginning of the project. Various modeling techniques are usually applied and evaluated before selecting which ones are best suited for the particular problem at hand. As some modeling techniques require data in a specific form, it's quite common to go back to the data preparation phase at this stage. This is an example of the iterativeness of CRISP-DM and analytics in general.

  After evaluating a number of models, it's time to select a set of candidate models to be methodically assessed. The assessment should estimate the effectiveness of the results in terms of accuracy, as well as ease of use in terms of interpretation of the results. If the assessment shows that we have found models that meet the necessary criteria, it's time for a more thorough evaluation, otherwise the work on finding suitable models has to continue.

5.6.4.4.1 Predictive maintenance example

With the business goal of finding models that help to reduce downtime of pump stations and maintenance costs, as well as avoiding unnecessary maintenance work, we form the hypothesis that pumps with similar characteristics will follow a similar pattern in terms of maintenance needs, and that it's possible to reuse knowledge from prior cases to make decisions. To make use of this hypothesis and test it, we create three models.

- **Action Prediction Model:** The action records are used to create a classification model that can predict what actions to take given a certain set of input data, e.g. the pump is vibrating strongly, the water is corrosive, it has been 14 months since it was serviced, and given prior cases, replacing the bearing and lubricating the pump with oil are likely to be the best actions. A decision tree-based model is selected since the data is highly heterogeneous and contains many categorical values. An assessment shows that the best performing model is based on the Random Forests method.

- **Forecasting Model:** To be able to predict future failures and needed maintenance in advance, a forecasting model is applied to the historic vibration sensor measurements. Two models, one based on the ARIMA (Autoregressive Integrated Moving Average) method, and the other on the ETS (Error—Trend—Seasonal) method performs well, and after assessment, the ETS-based model is selected.
- **Similar Pump Model:** To create a model that can be used to determine similarity between pumps, there exists a number of similarity and clustering techniques, such as k-nearest-neighbor and k-means. After some reasoning, it is decided to use a k-means-based model. These models require the number of clusters to be set as a parameter, and to determine the most appropriate number of clusters a decision tree is trained to classify which cluster a pump should belong to. The benefit of this is that trained pump maintenance experts are able to inspect the decision trees to determine which number of clusters produces the most realistic decision tree.

5.6.4.5 Evaluation

Now the project is nearing its end and it's time to evaluate the models from a business perspective using the success criteria that were defined at the beginning of the project. It is also customary to spend some time reviewing the project and draw conclusions about what was good and bad. This will be valuable input for future projects. At the end of the evaluation phase, a decision whether to deploy the results or not should be made.

5.6.4.5.1 Predictive maintenance example

To evaluate how well the models perform in the real world, a set of example cases are selected and the results are studied and verified by maintenance staff with several years of experience from working in the field. Several variations of the models, with slightly different parameter settings, are evaluated and studied. Especially the action prediction model is analyzed to find which version recommends the most correct actions compared to what the experts would recommend.

The two different approaches are evaluated:

- Action Forecasting: This approach proved to provide stable results that are easily interpreted by both humans and machines. A discussion was undertaken as to whether this could be used to automatically create work orders if the forecast was within certain bounds of confidence.
- Similar Pump Recommendations: This approach was much appreciated since it provided the staff with empirical data about how pumps under similar conditions have evolved, i.e. failed or been subjected to early maintenance.

A decision was made to deploy both approaches and combine them in one report.

5.6.4.6 Deployment

At this last phase in the project, the models are deployed and integrated into the organization. This can mean several things, such as writing a report to disseminate the results, or integrating the model into an automated system. This part of the project involves the customer directly, who has to provide the resources needed for an effective deployment.

The deployment phase also includes planning for how to monitor the models and evaluate when they have played out their role or need to be maintained.

As last steps, a final report and project review should be performed.

5.6.4.6.1 Predictive maintenance example

- Data from the pump stations is read automatically every day. A new batch job is deployed that is triggered when all readings have been collected. The batch job performs all the necessary data transformations and data loading needed before applying the models.
- A new routine was implemented that generates a 30-day forecast for all pumps, and then evaluates the action prediction model on each pump and its forecasted data.
- For those pumps that have actions predicted, a rule set is evaluated that checks the results against given thresholds to check for, for example, confidence of predicted actions. If these checks are positive, similar cases are retrieved and a report is generated and sent to the right people.
- The case retrieval method is implemented by first looking up the pumps that belong to the same cluster according to the similar pump model. For each similar pump, a scan is executed to find matches in time when the vibration characteristics were the same as the pump currently being evaluated. For each match, a lookup in the action records is performed to check what kind of actions and failure have

occurred within 30 days after the date of the match. The matches that have logged actions are then added to the report.

- Another batch job is also implemented, with the tasks of updating the pump and action records, as well as retraining the models periodically. This job is also responsible for evaluating the efficiency and correctness of the models as the system evolves over time. If the models seem to deteriorate, an administrator is notified.
- Finally, it's decided to inspect the system with the help of experts once a year to evaluate its performance.

5.7 Knowledge management

Section 5.6 covered analytics in the context of M2M. Here, we investigate Knowledge Management Frameworks. Firstly, we must look at the concept of knowledge, which in every day usage relates to information, understanding, or skill you get from experience or education. Within the context of ICT systems, the term "knowledge" mostly arises from the application of two other concepts: data and information, illustrated in Figure 5.16. We discuss here the relationships between these terms, and in the next section we discuss reference architecture for knowledge management within M2M and IoT solutions.

5.7.1 Data, information, and knowledge

For our purposes, we use the following definitions of data:

Data: Data refers to "unstructured facts and figures that have the least impact on the typical manager" (Thierauf 1999). With regards to IoT

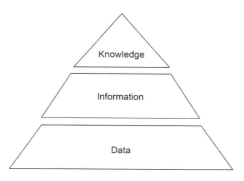

FIGURE 5.16

Data, Information, and Knowledge.

solutions, however, data includes both useful and irrelevant or redundant facts, and in order to become meaningful, needs to be processed.

Information: Within the context of IoT solutions, information is data that has been contextualized, categorized, calculated, and condensed (Davenport & Prusak 2000). This is where data has been carefully curated to provide relevance and purpose (Bali et al. 2009) for the decision-makers in question. The majority of ICT solutions can be viewed as either storing information or processing data to become information.

Knowledge: Knowledge, meanwhile, relates to the ability to understand the information presented, and using existing experience, the application of it within a certain decision-making context.

As we discussed in Section 5.6, data-driven scientific discovery is an important emerging paradigm for computing for IoT and cloud computing. So-called "big data" brings both opportunities and challenges ranging from capture, curation, storage, analysis, visualization, and search methods. The necessity to integrate across heterogeneous complex data resources in order to create information and, ultimately, knowledge, is one of the key research areas of our time.

For IoT solutions to be practicable, the data management and information presentation within them needs to take into consideration real-time performance, complexity, and the human-data interface. Knowledge management in this context needs to perform a careful balancing act between the sheer speed of incoming data sets and the provision of a user-centric presentation view. Due to the nature of big data, as we discussed in previous sections, two key issues emerge:

- Managing and storing the temporal knowledge created by IoT solutions. IoT solutions data will evolve rapidly over time, the temporal nature of the "knowledge" as understood at a particular point in time will have large implications for the overall industry. For example, it could affect insurance claims if the level of knowledge provided by an IoT system could be proven to be inadequate.
- Life-cycle management of knowledge within IoT systems. Closely related to analytics, the necessity to have a lifecycle plan for the data within a system is a strong requirement.

Having covered the differences between data, information, and knowledge, we now move to outlining a reference architecture for knowledge management in IoT solutions. Existing knowledge management frameworks have previously focused on clearly structured data, generally found in databases that can be stored in a form that is easily analyzed via various

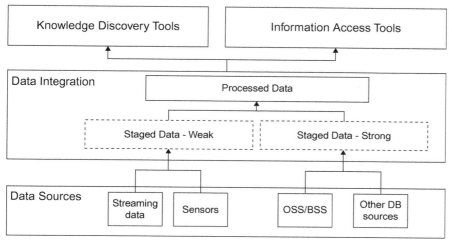

FIGURE 5.17

Knowledge Reference Architecture for M2M and IoT.

well-established tools. With sensor data and similar types, however, it is just as important to understand how to organize data effectively in order to extract knowledge from it.

5.7.2 A knowledge management reference architecture

Figure 5.17 outlines a high-level knowledge management reference architecture that illustrates how data sources from M2M and IoT may be combined with other types of data, for example, from databases or even OSS/BSS data from MNOs. There are three levels to the diagram: (1) data sources, (2) data integration, and (3) knowledge discovery and information access.

5.7.2.1 Data sources
Data sources refer to the broad variety of sources that may now be available to build enterprise solutions. There are several types of such data, which have been discussed in detail in Sections 5.5 and 5.6.

5.7.2.2 Data integration
The data integration layer allows data from different formats to be put together in a manner that can be used by the information access and knowledge discovery tools.

Staged Data: Staged data is data that has been abstracted to manage the rate at which it is received by the analysis platform. Essentially, "staged data" allows the correct flow of data to reach information access and knowledge discovery tools to be retrieved at the correct time. Big data and M2M analytics were discussed in detail in Section 5.5; here we focus on the data types required for staging the data appropriately for knowledge frameworks. There are two main types of data: weak data and strong data. This definition is in order to differentiate between the manner in which data is encoded and its contents — for example, the difference between XML and free text.

Strong Type Data: Strong type data refers to data that is stored in traditional database formats, i.e. it can be extracted into tabular format and can be subjected to traditional database analysis techniques. Strong data types often have the analysis defined beforehand, e.g. by SQL queries written by developers towards a database.

Weak Type Data: Weak type data is data that is not well structured according to traditional database techniques. Examples are streaming data or data from sensors. Often, this sort of data has a different analysis technique compared to strong type data. In this case, it may be that the data itself defines the nature of the query, rather than being defined by developers and created in advance. This may allow insights to be identified earlier than in strong type data.

5.7.2.3 Processed data

Processed data is combined data from both strong and weak typed data that has been combined within an IoT context to create maximum value for the enterprise in question. There are various means by which to do this processing — from stripping data separately and creating relational tables from it or pooling relevant data together in one combined database for structured queries. Examples could include combining the data from people as they move around the city from an operator's business support system with sensor data from various buildings in the city. A health service could then be created analyzing the end-users routes through a city and their overall health — such a system may be used to more deeply assess the role that air pollution may play in health factors of the overall population.

5.7.3 Retrieval layer

Once data has been collated and processed, it is time to develop insights from the data via retrieval. This can be of two main forms: Information Access and Knowledge Discovery.

5.7.3.1 Information access tools

Information access relates to more traditional access techniques involving the creation of standardized reports from the collation of strong and weak typed data. Information access essentially involves displaying the data in a form that is easily understandable and readable by end users. A variety of information access tools exist, from SQL visualization to more advanced visualization tools.

5.7.3.2 Knowledge discovery tools

Knowledge Discovery, meanwhile, involves the more detailed use of ICT in order to create knowledge, rather than just information, from the data in question. Knowledge Discovery means that decisions may be able to be taken on such outputs — for example, in the case where actuators (rather than just sensors) are involved, Knowledge Discovery Systems may be able to raise an alert that a bridge or flood control system may need to be activated.

In this chapter, we covered the basic technology building blocks that form an IoT or M2M solution. In the next chapters we cover the detailed technical architecture within which these building blocks can be placed.

IoT Architecture – State of the Art

CHAPTER OUTLINE

6.1 Introduction

In Chapter 5, we outlined the technology building blocks that form the basis of M2M and IoT solutions. Chapters 6, 7, and 8 provide a detailed conceptual overview of an M2M/IoT reference architecture introduced in Chapter 4. Chapters 7 and 8 also extend it with a description of a reference model. The term "Architecture Reference Model" (ARM) is borrowed from the IoT Architecture (IoT-A) European research project because the objective of the next three chapters is quite similar to that of the IoT-A ARM (Carrez et al. 2013). In other words, these chapters attempt to describe a combination of a reference model and a reference architecture. A reference model is a model that describes the main conceptual entities and how they are related to each other, while the reference architecture aims at describing the main functional components of a system as well as how the system works, how the system is deployed, what information the system processes, etc.

From Machine-to-Machine to the Internet of Things: Introduction to a New Age of Intelligence.
DOI: http://dx.doi.org/10.1016/B978-0-12-407684-6.00006-1
© 2014 Elsevier Ltd. All rights reserved.

An ARM is useful as a tool that establishes a common language across all the possible stakeholders of an M2M or IoT system. It can also serve as a starting point for creating concrete architectures of real systems when the relevant boundary conditions have been applied, for example, stakeholder requirements, design constraints, and design principles.

The approach followed in this chapter is to present the state-of-the-art of M2M reference models/high-level architectures/ARMs as well as the corresponding state-of-the-art in IoT frameworks. We then present the an adapted version of the IoT-A reference model (Chapter 7) and Reference Architecture (Chapter 8). In Part III, each specific use case describes a real system with a specific architecture accompanied with references to the ARM.

6.2 State of the art

Several Reference Architectures and Models exist both for M2M and IoT systems. We choose to present the four most popular ones from which the specific Reference Architecture and Model of this book have borrowed concepts and functions. The M2M system architectures are naturally more communication-oriented, while the IoT-related reference architectures and models are more holistic in their scope.

6.2.1 European Telecommunications Standards Institute M2M/oneM2M

The European Telecommunications Standards Institute (ETSI) in 2009 formed a Technical Committee (TC) on M2M topics aimed at producing a set of standards for communication among machines from an end-to-end viewpoint. The technical committee consisted of representatives from telecom network operators, equipment vendors, administrations, research bodies, and specialist companies. The ETSI M2M specifications are based on specifications from ETSI as well as other standardization bodies such as the IETF (Internet Engineering Task Force), 3GPP (3rd Generation Partnership Project), OMA (Open Mobile Alliance), and BBF (Broadband Forum). ETSI M2M produced the first release of the M2M standards in early 2012, while in the middle of 2012 seven of the leading Information and Communications Technology (ICT) standards organizations (ARIB, TTC, ATIS, TIA, CCSA, ETSI, TTA) formed a global organization called oneM2M Partnership Project (oneM2M) in order to develop M2M specifications, promote the M2M business, and ensure the global functionality of M2M systems. During the writing of this book, the oneM2M specifications are still in progress, and therefore only the ETSI M2M specifications are described in this chapter.

6.2.1.1 ETSI M2M high-level architecture

Figure 6.1 shows the high-level ETSI M2M architecture.

This high-level architecture is a combination of both a functional and topological view showing some functional groups (FG) clearly associated with pieces of physical infrastructure (e.g. M2M Devices, Gateways) while other functional groups lack specific topological placement. There are two main domains, a network domain and a device and gateway domain. The boundary between these conceptually separated domains is the topological border between the physical devices and gateways and the physical communication infrastructure (Access network).

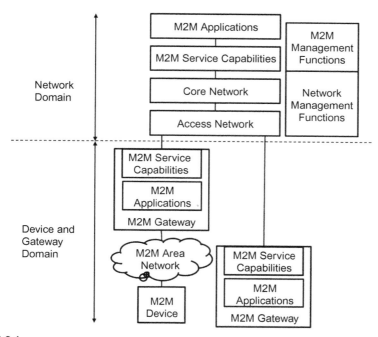

FIGURE 6.1

ETSI M2M High-Level Architecture.

(ETSI M2M TC 2013a)

The Device and Gateway Domain contains the following functional/ topological entities:

- **M2M Device**: This is the device of interest for an M2M scenario, for example, a device with a temperature sensor. An M2M Device contains M2M Applications and M2M Service Capabilities. An M2M device connects to the Network Domain either directly or through an M2M Gateway:
  - Direct connection: The M2M Device is capable of performing registration, authentication, authorization, management, and provisioning to the Network Domain. Direct connection also means that the M2M device contains the appropriate physical layer to be able to communicate with the Access Network.
  - Through one or more M2M Gateway: This is the case when the M2M device does not have the appropriate physical layer, compatible with the Access Network technology, and therefore it needs a network domain proxy. Moreover, a number of M2M devices may form their own local M2M Area Network that typically employs a different networking technology from the Access Network. The M2M Gateway acts as a proxy for the Network Domain and performs the procedures of authentication, authorization, management, and provisioning. An M2M Device could connect through multiple M2M Gateways.
- **M2M Area Network**: This is typically a local area network (LAN) or a Personal Area Network (PAN) and provides connectivity between M2M Devices and M2M Gateways. Typical networking technologies are IEEE 802.15.1 (Bluetooth), IEEE 802.15.4 (ZigBee, IETF 6LoWPAN/ROLL/ CoRE), MBUS, KNX (wired or wireless) PLC, etc.
- **M2M Gateway**: The device that provides connectivity for M2M Devices in an M2M Area Network towards the Network Domain. The M2M Gateway contains M2M Applications and M2M Service Capabilities. The M2M Gateway may also provide services to other legacy devices that are not visible to the Network Domain.

The Network Domain contains the following functional/topological entities:

- **Access Network**: this is the network that allows the devices in the Device and Gateway Domain to communicate with the Core Network. Example Access Network Technologies are fixed (xDSL, HFC) and wireless (Satellite, GERAN, UTRAN, E-UTRAN W-LAN, WiMAX).

- **Core Network**: Examples of Core Networks are 3GPP Core Network and ETSI TISPAN Core Network. It provides the following functions:
  - IP connectivity.
  - Service and Network control.
  - Interconnection with other networks.
  - Roaming.
- **M2M Service Capabilities**: These are functions exposed to different M2M Applications through a set of open interfaces. These functions use underlying Core Network functions, and their objective is to abstract the network functions for the sake of simpler applications. More details about the specific service capabilities are provided later in the chapter.
- **M2M Applications**: These are the specific M2M applications (e.g. smart metering) that utilize the M2M Service Capabilities through the open interfaces.
- **Network Management Functions**: These are all the necessary functions to manage the Access and Core Network (e.g. Provisioning, Fault Management, etc.).
- **M2M Management Functions**: These are the necessary functions required to manage the M2M Service Capabilities on the Network Domain while the management of an M2M Device or Gateway is performed by specific M2M Service Capabilities. There are two M2M Management functions:
  - M2M Service Bootstrap Function (MSBF): The MSBF facilitates the bootstrapping of permanent M2M service layer security credentials in the M2M Device or Gateway and the M2M Service Capabilities in the Network Domain. In the Network Service Capabilities Layer, the Bootstrap procedures perform, among other procedures, provisioning of an M2M Root Key (secret key) to the M2M Device or Gateway and the M2M Authentication Server (MAS).
  - M2M Authentication Server (MAS): This is the safe execution environment where permanent security credentials such as the M2M Root Key are stored. Any security credentials established on the M2M Device or Gateway are stored in a secure environment such as a trusted platform module.

An important observation regarding the ETSI M2M functional architecture is that it focuses on the high-level specification of functionalities within the M2M Service Capabilities functional groups and the open interfaces between the most relevant entities, while avoiding specifying in detail the internals of M2M Service Capabilities. However, the interfaces are specified in different levels of detail, from abstract to a specific mapping of an interface to a specific protocol (e.g. HTTP (Fielding 2000), IETF CoAP (Shelby et al. 2013a)). The most relevant entities in the ETSI M2M architecture are the M2M Nodes and M2M Applications. An M2M Node can be a Device M2M, Gateway M2M, or Network M2M Node (Figure 6.2). An M2M Node is a logical representation of the functions on an M2M Device, Gateway, and Network that should at least include a Service Capability Layer (SCL) functional group.

An M2M Application is the main application logic that uses the Service Capabilities to achieve the M2M system requirements. The application logic can be deployed on a Device (Device Application, DA), Gateway (Gateway Application, GA) or Network (Network Application, NA).

The SCL is a collection of functions that are exposed through the open interfaces or reference points mIa, dIa, and mId (ETSI M2M TC 2013b). Because the main topological entities that SCL can deploy are the Device, Gateway, and Network Domain, there are three types of SCL: DSCL (Device Service Capabilities Layer), GSCL (Gateway Service Capabilities Layer), and NSCL (Network Service Capabilities Layer). SCL functions utilize underlying networking capabilities through technology-specific interfaces. For example, an NSCL using a 3GPP type of access network uses 3GPP communication services interfaces. The ETSI M2M Service

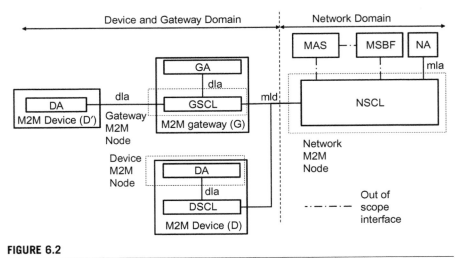

FIGURE 6.2

M2M Service Capabilities, M2M Nodes and Open Interfaces.

(ETSI M2M TC 2013a).

Capabilities are recommendations of functional groups for building SCLs, but their implementation is not mandatory, while the implementation of the interfaces mIa, dIa, and mId is mandatory for a compliant system. It is worth repeating that from the point of view of the ETSI M2M architecture, an M2M device can be either capable of supporting the mId interface (towards the NSCL) or the dIa interface (towards the GSCL). The specification actually distinguishes these two types of devices device D and device D' (D prime), respectively.

6.2.1.2 ETSI M2M service capabilities

All the possible Service Capabilities (where "x" is N(etwork), G(ateway), and D(evice)) are shown in Figure 6.3:

1. Application Enablement (xAE). The xAE service capability is an application facing functionality and typically provides the implementation of the respective interface: NAE implements the mIa interface and the GAE and DAE implement the dIa interface. The xAE includes registration of applications (xA) to the respective xSCL; for example, a Network Application towards the NSCL.

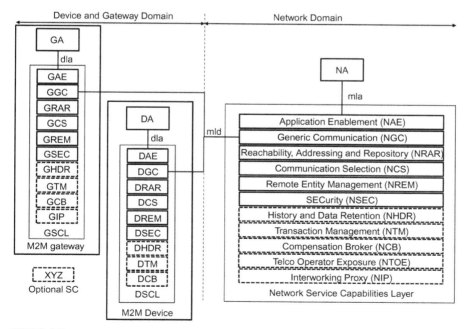

FIGURE 6.3

M2M Capabilities for different M2M Nodes.

In certain configurations xAE enables xAs to exchange messages to each other; for example, multiple Device Applications associated with the same M2M Gateway can exchange messages through the GAE. In certain configurations security operations such as authentication and authorization of applications is also performed by xAE.

2. Generic Communication (xGC). The NGC is the single point of contact for communication towards the GSCL and DSCL. It provides transport session establishment and negotiation of security mechanisms, potentially secure transmission of messages, and reporting of errors such as transmission errors. The GSC/DSC is the single point of contact for communication with the NSCL, and they both perform similar operations to the NGC (e.g. secure message transmissions to NSCL). The GSC performs a few more functions such as relaying of messages to/from NSCL from/to other SCs in the GSCL, and handles name resolution for the requests within the M2M Area Network.

3. Reachability, Addressing, and Repository (xRAR). This is one of the main service capabilities of the ETSI M2M architecture. The NRAR hosts mappings of M2M Device and Gateway names to reachability information (routable address information such as IP address and reachability status of the device such as up or down), and scheduling information relating to reachability, such as whether an M2M Device is reachable between 10 and 11 o'clock. It provides group management (creation/update/deletion) for groups of M2M Devices and Gateways, stores application (DA, GA, NA) data, and manages subscriptions to these data, stores registration information for NA, GSCL, and DSCL, and manages events (subscription notifications). The GRAR provides similar functionality to the NRAR, such as maintaining mappings of the names of M2M Devices or groups to reachability information (routable addresses, reachability status, and reachability scheduling), storing DA, GA, NSCL registration information, storing DA, GA, NA, GSCL, NSCL data and managing subscriptions about them, managing groups of M2M Devices, and managing events. Similar to NRAR and GRAR, the DRAR stores DA, GA, NA, DSCL, and NSCL data and manages subscriptions about these data, stores DA registration and NSCL information, provides group management for groups of M2M Devices and event management.

4. Communication Selection (xCS): This capability allows each xSCL to select the best possible communication network when there is more than one choice or when the current choice becomes unavailable due to communication errors. The NCS provides such a selection

mechanism based on policies for reaching an M2M Device or Gateway, while the GCS/DCS provides a similar selection mechanism for reaching the NSCL.

5. Remote Entity Management (xREM). The NREM provides management capabilities such as Configuration Management (CM) for M2M Devices and Gateways (e.g. installs management objects in device and gateways), collects performance management (PM) and Fault Management (FM) data and provides them to NAs or M2M Management Functions, performs device management to M2M Devices and Gateways such as firmware and software (application, SCL software) updates, device configuration, and M2M Area Network configuration. The GREM acts as a management client for performing management operations to devices using the DREM and a remote proxy for NREM to perform management operations to M2M Devices in the M2M Area Network. Examples of proxy operations are mediation of NREM-initiated software updates, and handling management data flows from NREM to sleeping M2M Devices. The DREM provides the CM, PM, and FM counterpart on the device (e.g. start collecting radio link performance data) and provides the device-side software and firmware update support.

6. SECurity (xSEC). These capabilities provide security mechanisms such as M2M Service Bootstrap, key management, mutual authentication, and key agreement (NSEC performs mutual authentication and key agreement while the GSEC and DESC initiate the procedures), and potential platform integrity mechanisms.

7. History and Data Retention (xHDR). The xHDR capabilities are optional capabilities, in other words, they are deployed when required by operator policies. These capabilities provide data retention support to other xSCL capabilities (which data to retain) as well as messages exchanged over the respective reference points.

8. Transaction Management (xTM). This set of capabilities is optional and provides support for atomic transactions of multiple operations. An atomic transaction involves three steps: (a) propagation of a request to a number of recipients, (b) collection of responses, and (c) commitment or roll back whether all the transactions successfully completed or not.

9. Compensation Broker (xCB). This capability is optional and provides support for brokering M2M-related requests and compensation between a Customer and a Service Provider. In this context a Customer and a Service Provider is an M2M Application.

10. Telco Operator Exposure (NTOE). This is also an optional capability and provides exposure of the Core Network service offered by a Telecom Network Operator.

11. Interworking Proxy (xIP). This capability is an optional capability and provides mechanisms for connecting non-ETSI M2M Devices and Gateways to ETSI SCLs. NIP provides mechanisms for non-ETSI M2M Devices and Gateways to connect to NSCL while GIP provides the functionality for non-compliant M2M Devices to connect to GSCL via the reference point dIa, and the DIP provides the necessary mechanisms to connect non-compliant devices to DSCL via the dIa reference point.

6.2.1.3 ETSI M2M interfaces

The main interfaces mIa, dIa, and mId (ETSI M2M TC 2013b) can be briefly described as follows:

- mIa: This is the interface between a Network Application and the Network Service Capabilities Layer (NSCL). The procedures supported by this interface are (among others) registration of a Network Application to the NSCL, request to read/write information to NSCL, GSCL, or DSCL, request for device management actions (e.g. software updates), subscription and notification of specific events.
- dIa: This is the interface between a Device Application and (D/G)SCL or a Gateway Application and the GSCL. The procedures supported by this interface are (among others) registration of a Device/Gateway Application to the GSCL, registration of a Device Application to the DSCL, request to read/write information to NSCL, GSCL, or DSCL, subscription and notification of specific events.
- mId: This is the interface between the Network Service Capabilities Layer (NSCL) and the GSCL or the DSCL. The procedures supported by this interface are (among others) registration of a Device/Gateway SCL to the NSCL, request to read/write information to NSCL, GSCL, or DSCL, subscription and notification of specific events.

6.2.1.4 ETSI M2M resource management

The ETSI M2M architecture assumes that applications (DA, GA, NA) exchange information with SCLs by performing CRUD (Create, Read, Update, Delete) operations on a number of Resources following the RESTful (Representational State Transfer) architecture paradigm (Fielding 2000). One of the principles of this paradigm is that representations of uniquely addressed resources are transferred from the entity that hosts these resources to the requesting entity. In the ETSI M2M architecture, all the state information maintained in the SCLs is modeled as a resource structure that architectural entities operate on. A very simplified view of the resource structure is a collection of containers of information structured in hierarchical manner following a corresponding hierarchy of a unique naming structure. In addition to the CRUD operations, ETSI M2M defines two more operations: NOTIFY and EXECUTE. The NOTIFY operation is triggered upon a change in the representation of a resource, and results in a notification sent to the entity that originally subscribed to monitor changes to the resource in question. This operation is not an orthogonal operation to the CRUD set, but can be implemented by an UPDATE operation from the resource host towards the requesting entity. The EXECUTE operation is not orthogonal as well, but can be implemented by an UPDATE operation with no parameters from the requesting entity to a specific resource. When a requesting entity issues an EXECUTE operation towards a specific resource, the specific resource executes a specific task.

The following example in Figure 6.4 demonstrates how an ETSI M2M entity communicates with another entity using the CRUD and NOTIFY operations. Assume that a device application (DA) is programmed to send a sensor measurement to a network application (NA). The DA using the DSCL updates the representation of a specific resource (R_a) residing on the NSCL (steps 1 and 2 in Figure 6.4). The NA has configured the NSCL to be notified when the specific resource is updated, in which case the NA reads the updated representation (steps 4 and 5, Figure 6.4).

The root of the hierarchical resource tree is the <sclBase> resource, and contains all the other resources hosted by the SCL. The root has a unique identifier. In case the RESTful architecture is implemented in a real system by using web resources, the <sclBase> has an absolute URI (Universal Resource Identifier), for example, "`http://m2m.operator1.com/ some/path/to/base`". The top-level structure of the <sclBase> resource is shown in Figure 6.5. The different fonts used in this figure denote

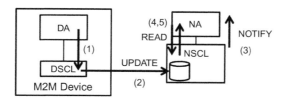

FIGURE 6.4

Communication Between DA and NA Using the SCLs.

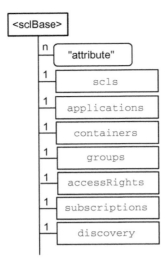

FIGURE 6.5

The Top-Level Structure of the <sclBase> Resource.

different information semantics. A term between the symbols "<" and ">" denotes an arbitrary resource name; for example, <sclBase> in Figure 6.5. A term within quotes ("") denotes a placeholder for one or more fixed names. In this specific case, "attribute" represents a member of a fixed list of attributes for the resource <sclBase>. A term annotated in Courier New font such as scls denotes a literal resource name used by the specification as is. The <sclBase> is structured as a tree with different branches, each of which is annotated with a designation of its cardinality. For example, the <sclBase> contains *n* attributes, one scls resource, one applications resource, etc. For more information about the resource structure of ETSI M2M specification, please refer to (ETSI M2M TC, 2013a).

The ETSI M2M specification also describes an HTTP (Fielding 2000) and a CoAP (Shelby et al., 2013a) binding for the RESTful resources stored in the SCLs, as well as for the implementation of the mId interface.

6.2.2 International Telecommunication Union — Telecommunication sector view

The Telecommunication sector of the International Telecommunication Union (ITU-T) has been active on IoT standardization since 2005 with the Joint Coordination Activity on Network Aspects of Identification Systems (JCA-NID), which was renamed to Joint Coordination Activity on IoT

(JCA-IoT) in 2011. During the same year apart from this coordination activity on IoT, ITU-T formed the specific IoT Global Standards Initiative (IoT-GSI) activity in order to address specific IoT-related issues. The latest ITU-T Recommendation, Y.2060 (ITU-T 2013) (Vermesan & Friess 2013), provides an overview of the IoT space with respect to ITU-T. This recommendation describes a high-level overview of the IoT domain model and the IoT functional model as a set of Service Capabilities similar to ETSI-M2M.

The ITU-T IoT domain model includes a set of physical devices that connect directly or through gateway devices to a communication network that allows them to exchange information with other devices, services, and applications. The physical world of things is reflected by an information world of virtual things that are digital representations of the physical things (not necessarily a one-to-one mapping because multiple virtual things can represent one physical thing). The devices in this model include mandatory communication capabilities and optional sensing, actuation, and processing capabilities in order to capture and transport information about the things.

Regarding the Service Capabilities (Figure 6.6), starting from the Application Layer the ITU-T IoT model considers this layer as the host of specific IoT applications (e.g. remote patient monitoring). The Service & Application Support Layer (otherwise known as Service Support and Application Support Layer) consists of generic service capabilities used by all IoT applications, such as data processing and data storage, and specific service capabilities tailored to specific application domains, such as e-health or telematics. The Network Layer provides networking capabilities such as Mobility Management, Authentication, Authorization, and Accounting (AAA), and Transport Capabilities such as connectivity for IoT service data. The Device Layer includes Device Capabilities and Gateway Capabilities. The Device Capabilities include, among others, the direct device interaction with the communication network and therefore the Network Layer Capabilities, the indirect interaction with the Network Layer Capabilities through Gateway Devices, any *ad hoc* networking capabilities, as well as low-power operation capabilities (e.g. capability to sleep and wakeup) that affect communications. The Gateway Device Capabilities include multiple protocol support and protocol conversion in order to bridge the Network Layer capabilities and the device communication capabilities. In terms of Management Capabilities, these include the typical FCAPS (Fault, Configuration, Accounting, Performance, Security) model of capabilities as well as device management (e.g. device

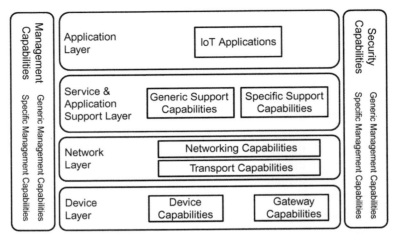

FIGURE 6.6

ITU-T IoT Reference Model.

(ITU-T 2013)

provisioning, software updates, activation/deactivation), network topology management (e.g. for local and short range networks), and traffic management. Specific management functionality related to a specific application domain is also included among the Management Capabilities. With respect to the Security Capabilities, this layer represents a grouping of different Security Capabilities required by other layers. The capabilities are grouped generically, such as AAA and message integrity/confidentiality support, and specifically, such as ones that are tailored to the specific application, e.g. mobile payment.

If the ITU-T and ETSI M2M models/architectures are compared, the two approaches are very similar in terms of Service Capabilities for M2M. However, the ITU-T IoT domain model with physical and virtual things, and the physical and virtual world model, shows the influences of more modern IoT architectural models and references such as the IoT-A (Carrez et al. 2013).

6.2.3 Internet Engineering Task Force architecture fragments

As of the writing of this book IETF has defined at least three working groups for addressing M2M and IoT, as was mentioned in Chapter 5: 6LoWPAN (IPv6 over Low-power WPAN), CoRE (Constrained RESTful Environments), and ROLL (Routing Over Low power and Lossy networks). The scope of the bulk of the specifications is shown in Figure 6.7, and it is clear that each set of specifications makes an attempt to address

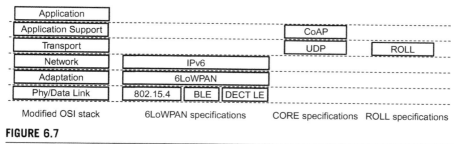

FIGURE 6.7

IETF Working Groups and Specification Scope.

a different part of the communication stack of a constrained device. The modified Open Systems Interconnection (OSI) model in which several layers are merged because of the implementation on constrained devices. This is illustrated in Figure 6.7 as one layer called Application Support which includes the Presentation and Session Layers combined. Moreover, one intermediate layer is introduced: the Adaptation Layer positioned between the Physical/Data Link and the Network Layer and whose main function is to adapt the Network Layer packets to Phy/Link layer packets among others. An example of an adaptation layer is the 6LoWPAN layer designed to adapt IPv6 packets to IEEE 8021.5.4/Bluetooth Low Energy (BLE)/DECT Low Energy packets. An example of an Application Support Layer is IETF Constrained Application Protocol (CoAP), which provides reliability and RESTful operation support to applications; however, it does not describe the specific names of resources a node should host. As seen earlier in Chapter 5, the IETF CoAP draft specification describes the Transport and Application Support Layers, which essentially defines the transport packet formats, reliability support on top of UDP, and a RESTful application protocol with GET/PUT/POST/DELETE methods similar to HTTP with CoAP clients operating on CoAP server resources. A CoAP server is just a logical protocol entity, and the name "server" does not necessarily imply that its functionality is deployed on a very powerful machine; a CoAP server can be hosted on a constrained device.

Apart from the core of the specifications, the IETF CoRE workgroup includes several other draft specifications that sketch parts of an architecture for IoT.

The CoRE Link Format specification (Shelby 2012) describes a discovery method for the CoAP resources of a CoAP server. For example, a CoAP client sending a request with the GET method to a specific well-defined server resource (`./well-known/core`) should receive a response with a list of CoAP resources and some of their capabilities (e.g. resource type, interface type). An accompanying draft specification, the CoRE interface specification (Shelby & Vial 2013), describes interface types and corresponding expected behavior of the RESTful methods (e.g. a sensor interface should support a GET method). Nevertheless, the response serialization (e.g. if the response is a temperature value in degrees Celsius) is not specified by these documents.

The IETF stack for IoT does not currently include any specifications that are similar to the profile specifications of other IoT technologies such as ZigBee (please refer to Chapter 5). By profile specification we mean a document that describes a list of profile names and their mappings to specific protocol stack behavior, specific information model, and specific serialization of this information model over the relevant communication medium. An example of a profile specification excerpt would mandate that an exemplary "Temperature" profile: (a) should support a resource called /temp, (b) the resource /temp must respond to a GET method request from a client, and (c) the response to a GET method request shall be a temperature value in degrees Celsius formatted as a text string with the format "<temperature value encoded in a decimal number >°C" (e.g "10°C"). It must be noted that the device profiles are used for ensuring interoperability between market products, and therefore it is not the responsibility of IETF to specify such details. Therefore, a step towards the specification of profiles was taken by the Internet Protocol for Smart Objects (IPSO) Alliance, which is mainly a market-promoting alliance. IPSO has published the IPSO Application Framework (Shelby & Chauvenet 2012) specification for this purpose.

The IETF CoRE working group has also produced a draft specification for a Resource Directory (Shelby et al. 2013b). A Resource Directory is a CoAP server resource (/rd) that maintains a list of resources, their corresponding server contact information (e.g. IP addresses or fully qualified domain name, or FQDN), their type, interface, and other information similar to the information that the CoRE Link Format document specifies (Figure 6.8a). An RD plays the role of a rendezvous mechanism for CoAP Server resource descriptions, in other words, for devices to publish the descriptions of the available resources and for CoAP clients to locate resources that satisfy certain criteria such as specific resource types (e.g. temperature sensor resource type).

While the Resource Directory is a rendezvous mechanism for CoAP Server resource descriptions, a Mirror Server (Vial 2012) is a rendezvous mechanism for CoAP Server resource presentations. A Mirror Server is a CoAP Server resource (/ms) that maintains a list of resources and their

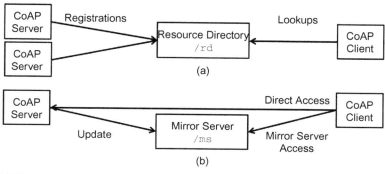

FIGURE 6.8

IETF CoRE Functional Components: (a) Resource Directory, (b) Mirror Server.

cached representations (Figure 6.8b). A CoAP Server registers its resources to the Mirror Server, and upon registration a new mirror server resource is created on the Mirror Server with a container (mirror representation) for the original server representation. The original CoAP Server updates the mirror representation either periodically or when the representation changes. A CoAP Client that retrieves the mirror representation receives the latest updated representation from the original CoAP Server. The Mirror Server is useful when the CoAP Server is not always available for direct access. An example of such a CoAP Server is one that resides on a real device whose communication capabilities are turned off in order to preserve energy, e.g. battery-powered radio devices whose radio and/or processor goes to sleep mode. Typically, a Mirror Server is hosted on a device or machine that is always available. Because CoAP as an application protocol is not yet widely deployed, while HTTP is ubiquitous, the IETF CoRE workgroup has included the fundamentals of a mapping process between HTTP and CoAP in the IETF CoAP specification as well as a set of guidelines for the interworking between HTTP and CoAP (Castellani et al. 2013) (Figure 6.9). The interworking issues appear when an HTTP Client accesses a CoAP Server through an HTTP-CoAP proxy or when a CoAP Client accesses an HTTP Server through a CoAP-HTTP proxy (Figure 6.9a). The mapping process is not straightforward for a number of reasons. The main is the different transport protocols used by the HTTP and CoAP: HTTP uses TCP while CoAP uses UDP. The guidelines focus more on the HTTP-to-CoAP proxy and recommend addressing schemes (e.g. how to map a CoAP resource address to an HTTP address), mapping between HTTP and CoAP response codes, mapping between different media types carried in the HTTP/CoAP payloads, etc. As an

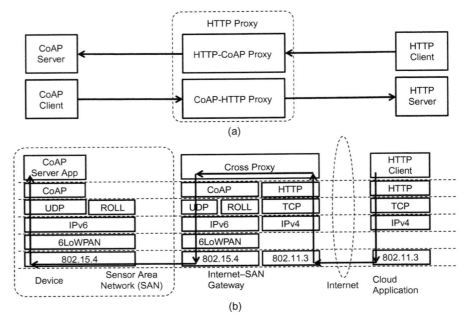

FIGURE 6.9

IETF CoRE HTTP Proxy: (a) possible configurations, (b) example layer interaction upon a request from a HTTP Client to a CoAP Server via a HTTP Proxy.

example, consider the case that an HTTP Client sends an HTTP request to a CoAP server (Figure 6.9a) through a Gateway Device hosting an HTTP-CoAP Cross Proxy. The Gateway Device connects to the Internet via an Ethernet cable using a LAN, and on the CoAP side the CoAP server resides on a Sensor/Actuator (SAN) based on the IEEE 802.15.4 PHY/MAC. The HTTP request needs to include two addresses, one for reaching the Cross Proxy and one for reaching the specific CoAP Server in the SAN. The default recommended address mapping is to append the CoAP resource address (e.g. `coap://s.coap.example.com/foo`) to the Cross proxy address (e.g. `http://p.example.com/.well-known/core/`), resulting in `http://p.example.com/.well-known/core/coap://s.coap.example.com/foo`. The request is in plain text format and contains the method (GET). It traverses the IPv4 stack of the client, reaches the gateway, traverses the IPv4 stack of the gateway and reaches the Cross proxy. The request is translated to a CoAP request (binary format) with a destination CoAP resource `coap://s.coap.example.com/foo`, and it is dispatched in the CoAP stack of the gateway, which sends it over the SAN to the end device. A response is sent from the end device and follows the reverse path in the

SAN in order to reach the gateway. The Cross proxy translates the CoAP response code to the corresponding HTTP code, transforms the included media, creates the HTTP response, and dispatches it to the HTTP client. While the described example scenario seems straightforward, in practice, the Cross proxy needs to handle all problematic situations and peculiarities of the CoAP and HTTP protocols, e.g. asynchronous behavior of the observe mode of CoAP (Hartke 2013). An interested reader can refer to the relevant specifications for further information.

6.2.4 **Open Geospatial Consortium architecture**

The Open Geospatial Consortium (OGC 2013) is an international industry consortium of a few hundred companies, government agencies, and universities that develops publicly available standards that provide geographical information support to the Web, and wireless and location-based services. OGC includes, among other working groups, the Sensor Web Enablement (SWE) (OGC SWE 2013) domain working group, which develops standards for sensor system models (e.g. Sensor Model Language, or SensorML), sensor information models (Observations & Measurements, or O&M), and sensor services that follow the Service-Oriented Architecture (SOA) paradigm, as is the case for all OGC-standardized services. The functionality that is targeted by OGC SWE includes:

- Discovery of sensor systems and observations that meet an application's criteria.
- Discovery of a sensor's capabilities and quality of measurements.
- Retrieval of real-time or time-series observations in standard encodings.
- Tasking of sensors to acquire observations.
- Subscription to, and publishing of, alerts to be issued by sensors or sensor services based upon certain criteria.

OGC SWE includes the following standards:

- SensorML and Transducer Model Language (TML), which include a model and an XML schema for describing sensor and actuator systems and processes; for example, a system that contains a temperature sensor measuring temperature in Celsius, which also involves a process for converting this measurement to a measurement with Fahrenheit units.
- Observations and Measurements (O&M), which is a model and an XML schema for describing the observations and measurements for a sensor (Observations and Measurements, O&M).

- SWE Common Data model for describing low-level data models (e.g. serialization in XML) in the messages exchanged between OGC SWE functional entities.
- Sensor Observation Service (SOS), which is a service for requesting, filtering, and retrieving observations and sensor system information. This is the intermediary between a client and an observation repository or near real-time sensor channel.
- Sensor Planning Service (SPS), which is a service for applications requesting a user-defined sensor observations and measurements acquisition. This is the intermediary between the application and a sensor collection system.
- PUCK, which defines a protocol for retrieving sensor metadata for serial port (RS232) or Ethernet-enabled sensor devices.

An example of how these standards relate to each other is shown in Figure 6.10. Because OGC follows the SOA paradigm, there is a registry (CAT) that maintains the descriptions of the existing OGC services, including the Sensor Observation and Sensor Planning Services. Upon installation the sensor system using the PUCK protocol retrieves the SensorML description of sensors and processes, and registers them with the Catalog so as to enable the discovery of the sensors and processes by client applications. The Sensor System also registers to the SOS and the SOS registers to the Catalog. A client application #1 requests from the Sensor Planning Service that the Sensor System be tasked to sample its

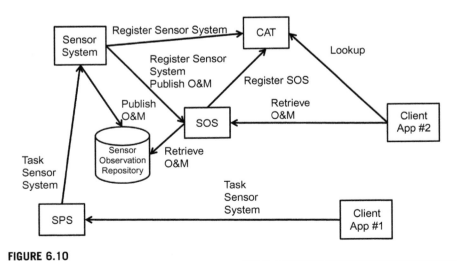

FIGURE 6.10

OGC functional architecture and interactions.

sensors every 10 seconds and publish the measurements using O&M and the SWE Common Data model to the SOS. Another client application #2 looks up the Catalog, aiming at locating an SOS for retrieving the measurements from the Sensor System. The application receives the contact information of the SOS and requests from the sensor observations from the specific sensor system from the SOS. As a response, the measurements from the sensor system using O&M and the SWE Common Data model are dispatched to the client application #2.

As can be seen from the description, the OGC SWE specifications are more information-centric than communication-centric, as are the ETSI M2M, ITU-T and IETF specifications. The main objective of the OGC standards is to enable data, information, and service interoperability.

This chapter provided an overview of the state-of-the-art within the reference architectures for M2M and IoT, including ETSI, ITU-T, IETF, and OGC. In Chapter 7, we now cover the ARM.

Architecture Reference Model

7.1 Introduction

This chapter provides an overview of the Architecture Reference Model (ARM) for IoT, including descriptions of the domain, information, and functional models. Chapter 8 then outlines the Reference Architecture.

From Machine-to-Machine to the Internet of Things: Introduction to a New Age of Intelligence.
DOI: http://dx.doi.org/10.1016/B978-0-12-407684-6.00007-3

7.2 Reference model and architecture

An ARM consists of two main parts: a Reference model and a Reference Architecture. For describing an IoT ARM, we have chosen to use the IoT-A ARM (Carrez et al. 2013) as a guide because it is currently the most complete model and reference. However, a real system may not have all the modeled entities or architectural elements described in this chapter, or it could contain other non-IoT-related entities. This chapter serves the purpose of modeling the IoT part of a whole system and does not try to propose an all-encompassing architecture.

The foundation of an IoT Reference Architecture description is an IoT reference model. A reference model describes the domain using a number of sub-models (Figure 7.1). The domain model of an architecture model captures the main concepts or entities in the domain in question, in this case M2M and IoT. When these common language references are established, the domain model adds descriptions about the relationship between the concepts. These concepts and relationships serve the basis for the

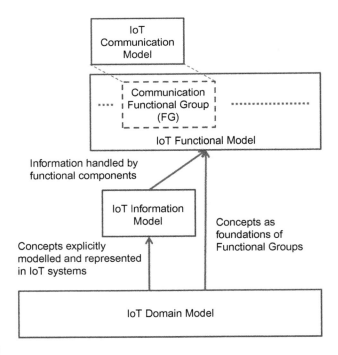

FIGURE 7.1

IoT Reference Model.

(Carrez et al. 2013)

development of an information model because a working system needs to capture and process information about its main entities and their interactions. A working system that captures and operates on the domain and information model contains concepts and entities of its own, and these need to be described in a separate model, the functional model. An M2M and IoT system contain communicating entities, and therefore the corresponding communication model needs to capture the communication interactions of these entities. These are a few examples of sub-models that we use in this chapter for the IoT reference model.

Apart from the reference model, the other main component of an ARM is the Reference Architecture. A System Architecture is a communication tool for different stakeholders of the system. Developers, component and system managers, partners, suppliers, and customers have different views of a single system based on their requirements and their specific interactions with the system. As a result, describing an architecture for M2M and IoT systems involves the presentation of the multiple facets of the systems in order to satisfy the different stakeholders (Kruchten 1995; Rozanski & Woods 2005, 2011). The task becomes more complex when the architecture to be described is on a higher level of abstraction compared with the architecture of real functioning systems. The high-level abstraction is called Reference Architecture as it serves as a reference for generating concrete architectures and actual systems, as shown in the Figure 7.2 (Muller & Hole 2007). Concrete architectures are instantiations of rather abstract and high-level Reference Architectures. A Reference Architecture captures the essential parts of an architecture, such as design principles, guidelines, and required parts (such as entities), to monitor and interact with the physical world for the case of an IoT Reference Architecture. A concrete architecture can be further elaborated and mapped into real world components by designing, building, engineering, and testing the different components of the actual system. As the figure implies, the whole process is iterative, which means that the actual deployed system in the field provides invaluable feedback with respect to the design and engineering choices, current constraints of the system, and potential future opportunities that are fed back to the concrete architectures. The general essentials out of multiple concrete architectures can then be aggregated, and contribute to the evolution of the Reference Architecture.

The IoT architecture model is related to the IoT Reference Architecture as shown in Figure 7.3. This figure shows two facets of the IoT ARM: (a) how to actually create an IoT ARM, and (b) how to use it with respect to building actual systems. In this chapter we mainly focus on how to use an

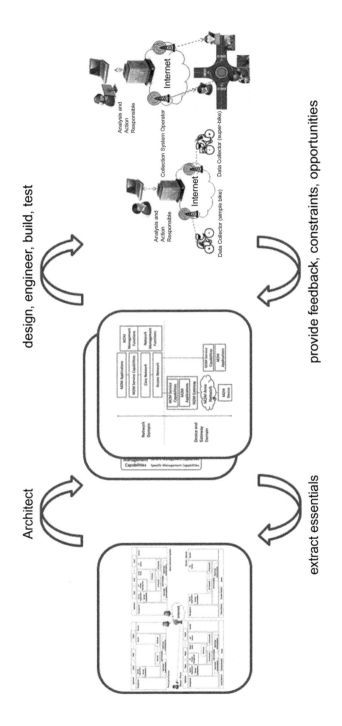

Architect

design, engineer, build, test

provide feedback, constraints, opportunities

extract essentials

Reference Architecture   Concrete Architecture(s)   Actual systems

FIGURE 7.2

From reference to concrete architectures and actual systems.

(Adapted from Muller & Hole 2007)

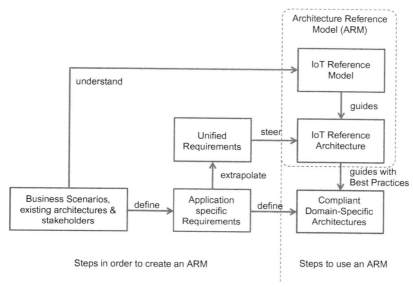

FIGURE 7.3

IoT Reference Model and Reference Architecture dependencies.

(Adapted from Carrez et al. 2013)

ARM; interested readers in the process of creation of an ARM are referred to the IoT-A ARM specification (Carrez et al. 2013). Moreover, the requirement collection and generation process is outlined in Pastor et al. (2011); the specific Unified Requirements (Figure 7.3) collected by the IoT-A exist online at (IoT-A 2013). The IoT reference model guides the process of creating an IoT Reference Architecture because it includes at least the IoT Domain model that impacts several architecture components as seen briefly earlier (e.g. functional groups), and as will be seen more extensively later in the book.

7.3 IoT reference model

7.3.1 IoT domain model

A domain model defines the main concepts of a specific area of interest, in this case the IoT. These concepts are expected to remain unchanged over the course of time, even if the details of an ARM may undergo continuous transformation or evolution over time. The domain model captures the basic attributes of the main concepts and the relationship between these concepts. A domain model also serves as a tool for human communication between people working in the domain in question and between people who work across different domains.

7.3.1.1 Model notation and semantics

For the purposes of the description of the domain model, we use the Unified Modeling Language (UML) (OMG 2013, NoMagic 2013) Class diagrams in order to present the relationships between the main concepts of the IoT domain model. The Class diagrams consist of boxes that represent the different classes of the model connected with each other through typically continuous lines or arrows, which represent relationships between the respective classes. Each class is a descriptor of a set of objects that have similar structure, behavior, and relationships. A class contains a name (e.g. Class A in Figure 7.4) and a set of attributes and operations. For the description of the IoT domain model, we will use only the class name and the class attributes, and omit the class operations. Notation-wise this is represented as a box with two compartments, one containing the class name and the other containing the attributes. However, for the IoT domain model description, the attribute compartment will be empty in order not to clutter the complete domain model. The relevant and interesting attributes will be described in the text instead.

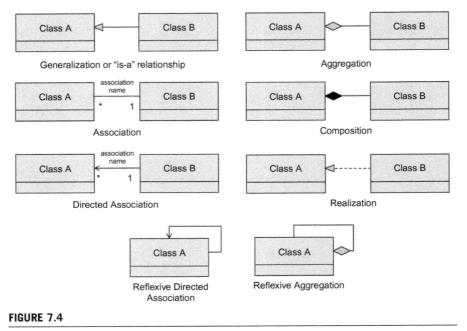

FIGURE 7.4

UML Class diagram main modeling concepts.

The following modeling relationships between classes (Figure 7.4) are needed for the description of the IoT Domain Model: Generalization/Specialization, Aggregation and Reflexive Aggregation, Composition, Directed Association and Reflexive Directed Association, and Realization.

The Generalization/Specialization relationship is represented by an arrow with a solid line and a hollow triangle head. Depending on the starting point of the arrow, the relationship can be viewed as a generalization or specialization. For example, in Figure 7.4, Class A is a general case of Class B or Class B is special case or specialization of Class A. Generalization is also called an "is-a" relationship. For example, in Figure 7.4 Class B "is-a" Class A. A specialized class/subclass/child class inherits the attributes and the operations from the general/super/parent class, respectively, and also contains its own attributes and operations.

The Aggregation relationship is represented by a line with a hollow diamond in one end and represents a whole-part relationship or a containment relationship and is often called a "has-a" relationship. The class that touches the hollow diamond is the whole class while the other class is the part class. For example, in Figure 7.4, class B represents a part of the whole Class A, or in other words, an object of Class A "contains" or "has-a" object of Class B. When the line with the hollow diamond starts and ends in the same class, then this relationship of one class to itself is called Reflexive Aggregation, and it denotes that objects of a class (e.g. Class A in Figure 7.4) contain objects of the same class.

The Composition relationship is represented by a line with a solid black diamond in one end, and also represents a whole-part relationship or a containment relationship. The class that touches the solid black diamond is the whole class while the other class is the part class. For example, in Figure 7.4, Class B is part of Class A. Composition and Aggregation are very similar, with the difference being the coincident lifetime to the objects of classes related with composition. In other words, if an object of Class B is part of an object of Class A (composition), when the object of Class A disappears, the object of Class B also disappears.

A plain line without arrowheads or diamonds represents the Association relationship. However, in the presentation of the IoT Domain model, we will only use the Directed Association that is represented with a line with a normal arrowhead. While all the previous relationships have implicit names represented by additional symbols (hollow triangle head, diamonds), an Association (Directed or not) contains an explicit association name. The Directed Association implies navigability from a Class B to a Class A in Figure 7.4. Navigability means that objects of Class B have the necessary attributes to know that they relate to objects of Class A while the reverse is not true: objects of Class A can exist without having references to objects of Class B. When the arrow starts and ends at the same class, then the class is associated to itself with a Reflexive Directed Association, which means that an object of this class (e.g. Class A in Figure 7.4) is associated with objects of the same class with the specific named association.

An arrow with a hollow triangle head and a dashed line represents the Realization relationship. This relationship represents a association between the class that specifies the functionality and the class that realizes the functionality. For example, Class A in Figure 7.4 specifies the functionality while Class B realizes it.

Aggregations, Reflexive Aggregations, Associations (Directed or not) and Reflexive Associations (Directed or not) may contain multiplicity information such as numbers (e.g "1"), ranges (e.g. "0−1", open ranges "1...*"), etc. in one or the other end of the relationship line/arrow. These multiplicities denote the potential number of class objects that are related to the other class object. For example, in Figure 7.4, a plain association called "association name," relates one (1) object of Class B with zero (0) or more objects from Class A. An asterisk "*" denotes zero (0) or more.

7.3.1.2 Main concepts

The IoT is a support infrastructure for enabling objects and places in the physical world to have a corresponding representation in the digital world. The reason why we would like to represent the physical world in the digital world is to remotely monitor and interact with the physical world using software. Let's illustrate this concept with an example (Figure 7.5).

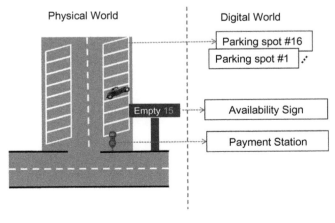

FIGURE 7.5

Physical vs. Virtual World.

Imagine that we are interested in monitoring a parking lot with 16 parking spots. The parking lot includes a payment station for drivers to pay for the parking spot after they park their cars. The parking lot also includes an electronic road sign on the side of the street that shows in real-time the number of empty spots. Frequent customers also download a smart phone application that informs them about the availability of a parking spot before they even drive on the street where the parking lot is located. In order to realize such a service, the relevant physical objects as well as their properties need to be captured and translated to digital objects such as variables, counters, or database objects so that software can operate on these objects and achieve the desired effect, i.e. detecting when someone parks without paying, informing drivers about the availability of parking spots, producing statistics about the average occupancy levels of the parking lot, etc. For these purposes, the parking lot as a place is instrumented with parking spot sensors (e.g. loops), and for each sensor, a digital representation is created (Parking spot #1−#16). In the digital world, a parking spot is a variable with a binary value ("available" or "occupied"). The parking lot payment station also needs to be represented in the digital world in order to check if a recently parked car owner actually paid the parking fee. Finally, the availability sign is represented to the digital world in order to allow notification to drivers that an empty lot is full for maintenance purposes, or even to allow maintenance personnel to detect when the sign is malfunctioning.

As seen from the example above, there is a fundamental difference between the IoT and today's Internet: today's Internet serves a rather virtual world of content and services (although these services are hosted on real physical machines), while IoT is all about interaction through the Internet with physical Things. M2M has a similar vision of representing unattended Devices accessible through a communication network in the digital world. Nevertheless, for the IoT model, the first class citizen is the Thing, and therefore Thing-oriented interaction is promoted as opposed to communication-oriented interaction for the M2M world.

As interaction with the physical world is the key for the IoT; it needs to be captured in the domain model (Figure 7.6). The first most fundamental interaction is between a human or an application with the physical world object or place. Therefore, a User and a Physical Entity are two concepts that belong to the domain model. A User can be a Human User, and the interaction can be physical (e.g. parking the car in the parking lot). The physical interaction is the result of the intention of the human to achieve a certain goal (e.g. park the car). In other occasions, a Human

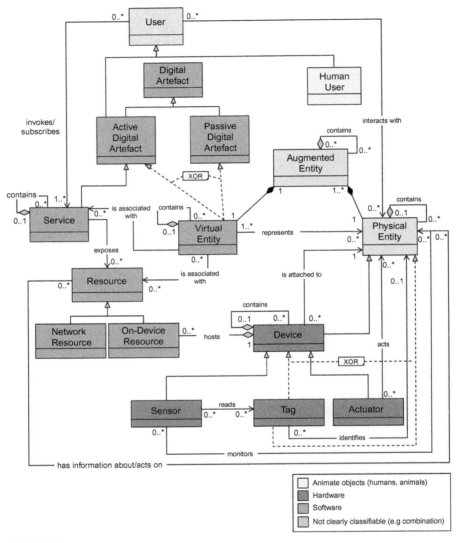

FIGURE 7.6

IoT Domain Model.

(Adapted from IoT-A 2013, Carrez et al. 2013)

User can also choose to interact with the physical environment by means of a service or application. This application is also a User in the domain model. A Physical Entity, as the model shows, can potentially contain other physical entities; for example, a building is made up of several floors, and each floor has several rooms.

The objects, places, and things represented as *Physical Entities* are the same as *Assets* mentioned earlier in the book. According to the Oxford Dictionary, an Asset "...is an item or property that is regarded as having value..."; therefore, the term *Asset* is more related to the business aspects of IoT. Because the domain model is a technical tool, we use the term *Physical Entity* instead of *Asset*.

A Physical Entity is represented in the digital world as a *Virtual Entity*. A Virtual Entity can be a database entry, a geographical model (mainly for places), an image or avatar, or any other *Digital Artifact*. One Physical Entity can be represented by multiple Virtual Entities, each serving a different purpose, e.g. a database entry of a parking spot denoting the spot availability, and an (empty/full) image of a parking spot on the monitor of the parking lot management system. Each Virtual Entity also has a unique identifier for making it addressable among other Digital Artifacts. A Virtual Entity representation contains several attributes that correspond to the Physical Entity current state (e.g. the parking spot availability). The Virtual Entity representation and the Physical Entity actual state should be synchronized whenever a User operates on one or the other, if of course that is physically possible. For example, a remotely controlled light (Physical Entity) represented by a memory location (Virtual Entity) in an application could be switched on/off by the User by changing the Virtual Entity representation, or in other words writing a value in the corresponding memory location. In this case, the real light should be turned on/off (Virtual to Physical Entity synchronization). On the other hand, if a Human User turns off the light by hand, then the Virtual Entity parameter that captures the state of the light should also be updated accordingly (Physical to Virtual Entity synchronization). There are cases that state synchronization can occur only one way. For example, the parking spot sensor representation in the digital world is updated whenever a car parks in the spot, but updating the digital representation does not mean that a car will magically land on that parking spot!

While discussing the concept of a Virtual Entity, we also introduced another concept, that of the Digital Artifact. A Digital Artifact is an artifact of the digital world, and can be passive (e.g. a database entry) or active (e.g. application software).

The model captures human-to-machine, application (active digital artifact)-to-machine, and M2M interaction when a digital artifact, and thus a User, interacts with a Device that is a Physical Entity. The model captures this special case of Devices being Physical and Virtual Entities as an Augmented Entity concept, which is a composition of the two constituent entities.

In order to monitor and interact with the Physical Entities through their corresponding virtual entities, the Physical Entities or their surrounding environment needs to be instrumented with certain kinds of Devices, or certain Devices need to be embedded/attached to the environment. The Devices are physical artifacts with which the physical and virtual worlds interact. Devices as mentioned before can also be Physical Entities for certain types of applications, such as management applications when the interesting entities of a system are the Devices themselves and not the surrounding environment. For the IoT Domain Model, three kinds of Device types are the most important:

1. **Sensors**: These are simple or complex Devices that typically involve a transducer that converts physical properties such as temperature into electrical signals. These Devices include the necessary conversion of analog electrical signals into digital signals, e.g. a voltage level to a 16-bit number, processing for simple calculations, potential storage for intermediate results, and potentially communication capabilities to transmit the digital representation of the physical property as well receive commands. A video camera can be another example of a complex sensor that could detect and recognize people.

2. **Actuators**: These are also simple or complex Devices that involve a transducer that converts electrical signals to a change in a physical property (e.g. turn on a switch or move a motor). These Devices also include potential communication capabilities, storage of intermediate commands, processing, and conversion of digital signals to analog electrical signals.

3. **Tags**: Tags in general identify the Physical Entity that they are attached to. In reality, tags can be Devices or Physical Entities but not both, as the domain model shows. An example of a Tag as a Device is a Radio Frequency Identification (RFID) tag, while a tag as a Physical Entity is a paper-printed immutable barcode or Quick Response (QR) code. Either electronic Devices or a paper-printed entity tag contains a unique identification that can be read by optical means (bar codes or QR codes) or radio signals (RFID tags). The reader Device operating on a tag is typically a sensor, and sometimes a sensor and an actuator combined in the case of writable RFID tags.

As shown in the model, Devices can be aggregation of other Devices e.g. a sensor node contains a temperature sensor, a Light Emitting Diode (LED, actuator), and a buzzer (actuator). Any type of IoT Device needs to (a) have energy reserves (e.g. a battery), or (b) be connected to the power grid, or (c) perform energy scavenging (e.g. converting solar radiation to energy). The Device communication, processing and storage, and energy reserve capabilities determine several design decisions such as if the resources should be on-Device or not, if the Device and therefore its resources and services will go into sleep mode or not, if the collected data can be saved locally or transmitted as soon as acquired, etc.

Resources are software components that provide data for, or are end-points for, controlling Physical Entities. Resources can be of two types, on-Device resources and Network Resources. An on-Device Resource is typically hosted on the Device itself and provides information, or is the control point for the Physical Entities that the Device itself is attached to. An example is a temperature sensor on a temperature node deployed in a room that hosts a software component that responds to queries about the temperature of the room. The Network Resources are software components hosted somewhere in the network or cloud. A Virtual Entity is associated with potentially several Resources that provide information or control of the Physical Entity represented by this Virtual Entity. Resources can be of several types: sensor resources that provide sensor data, actuator resources that provide actuation capabilities or actuator state (e.g. "on"/"off"), processing resources that get sensor data as input and provide processed data as output, storage resources that store data related to Physical Entities, and tag resources that provide identification data of Physical Entities.

Resources expose (monitor or control) functionality as Services with open and standardized interfaces, thus abstracting the potentially low-level implementation details of the resources. Services are therefore digital artifacts with which Users interact with Physical Entities through the Virtual Entities. Therefore, the Virtual Entities that are associated with Physical Entities instrumented with Devices that expose Resources are also associated with the corresponding resource Services. The associations between Virtual Entities and Services are such that a Virtual Entity may be monitored or controlled through potentially multiple redundant Resources or Services. Therefore, these associations are important to be maintained for lookup or discovery by the interested Users. It is important to note that IoT Services can be classified into three main classes according to their level of abstraction:

1. Resource-Level Services typically expose the functionality of a Device by exposing the on-Device Resources. In addition, these services typically handle quality aspects such as security, availability, and performance issues. Apart from the on-Device resources, there are also Network Resources hosted in more powerful machines or in the cloud that exposes Resource-Level Services and abstracts the location of the actual resources. An example of such a Network Resource is a

historical database of measurements of a specific resource on a specific Device. Resource-Level Services typically include interfaces that access resource information based on the identity of the resource itself.

2. Virtual Entity-Level Services provide information or interaction capabilities about Virtual Entities, and as a result the Service interfaces typically include an identity of the Virtual Entity.

3. Integrated Services are the compositions of Resource-Level and Virtual Entity-Level services, or any combination of both service classes.

An example of an instantiation of the IoT Domain Model is shown in Figure 7.7. For this instantiation, we use the example of a simple parking lot management system presented earlier, and we model only part of the real system. For example the part of the model that captures the Loop Sensor #21−#28 and the associated Physical and Virtual Entities is similar to the corresponding part of the model for the Loop Sensor #11−#18 and therefore omitted. We assume that each parking spot is instrumented with a metal sensing loop (Sensor), and half of the loops are physically wired to one sensor node (Device, Sensor Node #1) while the rest are wired to another sensor node (Device, Sensor Node #2). The sensor nodes may have different identifiers (e.g. Sensor #11−Sensor #18 for the first group, and

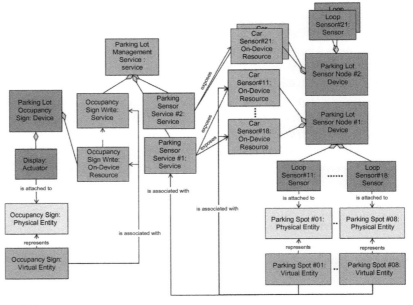

FIGURE 7.7

IoT Domain Model instantiation.

Sensor #21—Sensor #28 for the second group). The loop sensors can output different impedance based on the existence or the absence of a steel object. These impedances are translated by the sensor nodes into binary "0" and "1" readings. Each Parking Lot Sensor Node hosts as many Car Sensor Resources as assigned parking spots. There are also two Parking Sensor Services each running on a Sensor Node. The Parking Sensor Service #1 provides the reading of the Loop Sensor #11—#18, while the Parking Sensor Service #2 provides the readings of the Loop Sensor #21—#28. The Parking Lot Management Service has the necessary logic to map the sensor node readings to the appropriate occupancy indicator (e.g. "0" → "free", "1" → "occupied") and maps the parking spot sensor identifier to the corresponding parking spot identifier (e.g. Sensor #11—Sensor #18 → Spot #01—Spot #08, and Sensor #21—Sensor #28 → Spot #09—Spot #16). The Virtual Entities that represent the Parking Spot Physical Entities are database entries with the following attributes: (a) identity (e.g. ID #1—#16), (b) physical dimensions (e.g. 3m × 2m), (c) the location of the center of the rectangular spot with respect to the parking lot entrance (e.g. 3 meters to the west and 2 meter to the north), and (d) its occupancy level (e.g. "occupied" or "free"). The occupancy sign consists of a Device that contains a Display (Actuator) that is attached to a Physical Entity (the actual steel sign). The Device exposes one Resource with one Service that allows writing a value to the sign display. The actual steel sign (Physical Entity) is represented in the digital world with a Virtual Entity, which is a database entry with the following attributes: (a) Location of the sign (e.g. GPS location), (b) status (on/off), and (c) display value (e.g. 15 free spaces). The Parking Lot Management System is a composed Service that contains the parking spot occupancy services and the occupancy sign write service. Internally given the occupancy status of all the parking spots, it produces the total number of free spots and uses this attribute to update the actuator attached to the occupancy sign.

7.3.1.3 Further considerations

Identification of Physical Entities is important in an IoT system in order for any User to interact with the physical world though the digital world. Furness et al. (2009) describe two ways: (a) primary identification that uses natural features of a Physical Entity, and (b) secondary identification when using tags or labels attached to the Physical Entity. Both types of identification are modeled in the IoT Domain Model. Extracting natural features can be performed by a camera Device (Sensor) and relevant Resources that produce a set of features for specific Physical Entities. In addition, when it comes to physical spaces, a GPS Device or another type of location Device (e.g. an indoor location Device) can also be used to record the GPS coordinates of the space occupied by the Physical Entity. With respect to secondary identification, tags or labels attached to Physical Entities are modeled in the IoT Domain model, and there are relevant RFID or barcode technologies to realize such identification mechanisms.

Apart from identification, location and time information are important for the annotation of the information collected for specific Physical Entities and represented in Virtual Entities. Information without one or the other (i.e. location or time) is practically useless apart from the case of Body Area Networks (BAN, networks of sensors attached to a human body for live capture of vital signals, e.g. heart rate); that location is basically fixed and associated with the identification of the Human User. Nevertheless, in such cases, sometimes the location of the whole BAN or Human User is important for correlation purposes (e.g. upon moving outdoors, the Human User heart rate increases in order to compensate for the lower temperature than indoors). Therefore, the location, and often the timestamp of location, for the Virtual Entity can be modeled as an attribute of the Virtual Entity that could be obtained by location sensing resources (e.g. GPS or indoor location systems).

7.3.2 Information model

According to the Data−Information−Knowledge−Wisdom Pyramid (Rowley 2007), information is defined as the enrichment of data (raw values without relevant or usable context) with the right context, so that queries about who, what, where, and when can be answered. Because the Virtual Entity in the IoT Domain Model is the "Thing" in the Internet of Things, the IoT information model captures the details of a Virtual Entity-centric model.

Similar to the IoT Domain Model, the IoT Information Model is presented using Unified Modeling Language (UML) diagrams. As mentioned earlier, each class in a UML diagram contains zero or more attributes. These attributes are typically of simple types such as integers or text strings, and are represented with red text under the name of the class (e.g. entityType in the Virtual Entity class in Figure 7.8). A more complex attribute for a specific class A is represented as a class B, which is contained in class A with an aggregation relationship between class A and class B. Moreover, the UML diagram for describing the IoT Information Model contains additional notation not presented earlier. More specifically, the

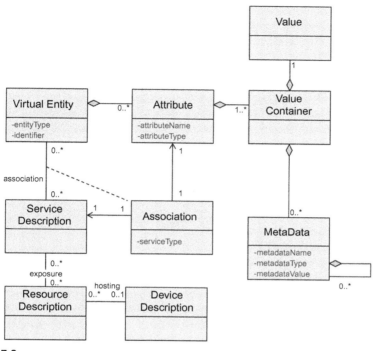

FIGURE 7.8

High-level IoT Information Model.

(Adapted from Carrez et al. 2013)

Association class in Figure 7.8 contains information about the specific association between a Virtual Entity and a related Service. In other words, while in the IoT Domain Model, one is interested in capturing the fact that a Virtual Entity and a Service are associated, and the IoT Information Model explicitly represents this association as part of the information maintained by an IoT system.

On a high-level, the IoT Information Model maintains the necessary information about Virtual Entities and their properties or attributes. These properties/attributes can be static or dynamic and enter into the system in various forms, e.g. by manual data entry or reading a sensor attached to the Virtual Entity. Virtual Entity attributes can also be digital synchronized copies of the state of an actuator as mentioned earlier: by updating the value of an Virtual Entity attribute, an action takes place in the physical world. In the presentation of the high-level IoT information model, we omit the attributes that are not updated by an IoT Device (sensor, tag) or the attributes that do not affect any IoT Device (actuator, tag), with the exception of essential attributes such as names and identifiers. Examples of omitted attributes that could exist in a real implementation are room names and floor numbers, in general, context information that is not directly related to IoT Devices, but that is nevertheless important for an actual system.

The IoT Information Model describes Virtual Entities and their attributes that have one or more values annotated with meta-information or metadata. The attribute values are updated as a result of the associated services to a Virtual Entity. The associated services, in turn, are related to Resources and Devices as seen from the IoT Domain Model. The IoT Information Model captures the above associations as follows.

A Virtual Entity object contains simple attributes/properties: (a) entityType to denote the type of entity, such as a human, car, or room (the entity type can be a reference to concepts of a domain ontology, e.g. a car ontology); (b) a unique identifier; and (c) zero or more complex attributes of the class Attributes. The class Attributes should not be confused with the simple attributes of each class. This class Attributes is used as a grouping mechanism for complex attributes of the Virtual Entity. Objects of the class Attributes, in turn, contain the simple attributes with the self-descriptive names attributeName and attributeType. As in the case of the entity type, the attribute type is the semantic type of the value (e.g. that the value is a temperature value), and can refer to an ontology such as the NASA quantities and units SWEET ontology (NASA JPL 2011). The Attribute class also contains a complex attribute ValueContainer that is a container of the multiple values that an attribute can take. The container includes complex attributes of the class Value and the class MetaData. The container contains exactly one value and meta-information (modeled as the class MetaData), such as a timestamp, describing this single value. Objects of the MetaData class can contain MetaData objects as complex attributes, as well as the simple attributes with the self-descriptive names metadataName, metadataType, and metadataValue.

As seen from the IoT Domain Model, a Virtual Entity is associated with Resources that expose Services about the specific Virtual Entity. This association between a Virtual Entity and its Services is captured in the Information Model with the explicit class called Association. Objects of this class capture the relationship between objects of the complex Attribute class (associated with a Virtual Entity) and objects of the Service Description class. The meaning of this explicit association is to link a specific attribute with the provider of the information or interaction functionality that is a Service associated with the Virtual Entity. Because the class Association describes the relationship between a Virtual Entity and

Service Description through the Attribute class, there is a dashed line between Association class and the line between the Virtual Entity and Service Description classes. The attribute serviceType can take two values: (a) "INFORMATION," if the associated service is a sensor service (i.e. allows reading of the sensor), or (b) "ACTUATION," if the associated service is an actuation service (i.e. allows an action executed on an actuator). In both cases, the eventual value of the attribute will be a result of either reading a sensor or controlling an actuator.

An example of an instantiation of the high-level information model is shown in Figure 7.9 following the parking lot example presented earlier. Here we don't show all the possible Virtual Entities, but only one corresponding to one parking spot. This Virtual Entity is described with one Attribute (among others) called hasOccupancy. This Attribute is associated with the Parking Lot Occupancy Service Description through the Occupancy Association. The Occupancy Association is the explicit expression of the association (line) between the Parking Spot #1 Virtual Entity and the Parking Lot Occupancy Service. Please note that the dashed arrows with hollow arrowheads represent the relationship "is instance of" for the information model, as opposed to the Realization relationship for the IoT Domain Model.

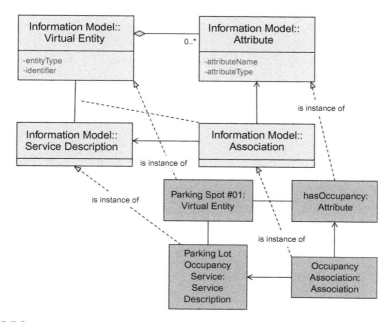

FIGURE 7.9

IoT Information Model example.

Throughout the description of the IoT Information Model, the reader might wonder about the mapping between the IoT Domain Model and the Information Model. Figure 7.10 presents the relationship between the core concepts of the IoT Domain Model and the IoT Information Model. The Information Model captures the Virtual Entity in the Domain Model being the "Thing" in the Internet of Things as several associated classes (Virtual Entity, Attribute, Value, MetaData, Value Container) that basically capture the description of a Virtual Entity and its context. The Device, Resource, and Service in the IoT Domain Model are also captured in the IoT Information Model because they are used as representations of the instruments and the digital interfaces for interaction with the Physical Entity associated with the Virtual Entity.

The IoT Information Model is by no means complete. As this is a description of an ARM, the Information Model is a very high-level model, and omits certain details that could potentially be required in a concrete architecture and an actual system. These details could be derived by specific requirements from specific use cases describing the target actual system. Because this chapter describes an ARM, we can only provide descriptions and guidelines about more information or models that could be used in real systems in conjunction with the proposed IoT Information Model.

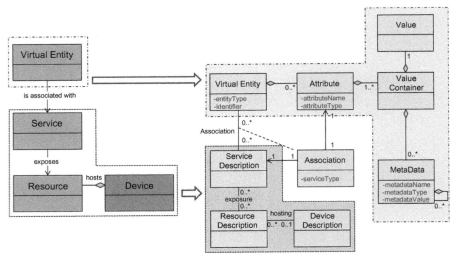

FIGURE 7.10

Relationship between core concepts of IoT Domain Model and IoT Information Model.

(Adapted from Carrez et al. 2013)

The Virtual Entity in the IoT Information Model is described with only a few simple attributes, the complex Attribute associated with sensor/actuator/tag services. As mentioned earlier, there are several other attributes or properties that could exist in a Virtual Entity description:

1. Location and its temporal information are important because Physical Entities represented by Virtual Entities exist in space and time. These properties are extremely important when the interested Physical Entities are mobile (e.g. a moving car). Even capturing the fact that the Physical Entity is static or mobile is also a piece of useful information. A mobile Physical Entity affects the associations between Attributes and related Services, e.g. a person moving close to a camera (sensor) is associated with the Device, Resource, and Services offered by the camera for as long as she stays within the field of view of the camera. In such cases, the temporal availability of the associations between Attributes and Services need to be captured, as availability denotes also temporal observability of the Virtual Entity.

2. Even non-moving Virtual Entities contain properties that are dynamic with time, and therefore their temporal variations need to be modeled and captured by an information model.

3. Information such as ownership is also important in commercial settings because it may determine access control rules or liability issues.

It is important to note that the Attribute class is general enough to capture all the interested properties of a Physical Entity, and thus provides an extensible model whose details can only be specified by the specific actual system in mind.

The Services in the IoT Domain Model are mapped to the Service Description in the IoT Information Model. The Service Description contains (among other information) the following (De et al. 2011, Martín et al. 2012):

1. Service type, which denotes the type of service, such as Big Web Service or RESTful Web Service. The interfaces of a service are described based on the description language for each service type, for example, Web Application Description Language (WADL) (Hadley 2009) for RESTful Web Services, Web Services Description Language (WSDL) (Christensen 2001) for Big Web Services, Universal Service Description Language (USDL) (Kadner & Oberle 2011), etc. The interface description includes, among other information, the invocation contact information, e.g. a Uniform Resource Locator (URL).

2. Service area and Service schedule are properties of Services used for specifying the geographical area of interest for a Service and the potential

temporal availability of a Service, respectively. For sensing services, the area of interest is equivalent to the observation area, whereas for actuation services the area of interest is the area of operation or impact.

3. Associated resources that the Service exposes.
4. Metadata or semantic information used mainly for service composition. This is information such as the indicator of which resource property is exposed as input or output, whether the execution of the service needs any conditions satisfied before invocation, and whether there are any effects of the service after invocation.

The IoT Information Model also contains Resource descriptions because Resources are associated with Services and Devices in the IoT Domain model. A Resource description contains the following information:

1. Resource name and identifier for facilitating resource discovery.
2. Resource type, which specifies if the resource is (a) a sensor resource, which provides sensor readings; (b) an actuator resource, which provides actuation capabilities (to affect the physical world) and actuator state; (c) a processor resource, which provides processing of sensor data and output of processed data; (d) a storage resource, which provides storage of data about a Physical Entity; and (e) a tag resource, which provides identification data for Physical Entities.
3. Free text attributes or tags used for capturing typical manual input such as "fire alarm, ceiling."
4. Indicator of whether the resource is an on-Device resource or network resource.
5. Location information about the Device that hosts this resource in case of an on-Device resource.
6. Associated Service information.
7. Associated Device description information.

A Device, as discussed in Chapter 5, is a Physical Entity that could have a sensor, actuator, or tag instantiation. An instantiation of a Device depends on the realization of it, and any information from dimensions of physical packaging to physical placement of sensors, actuators, tags, processors, memories, batteries, cables, etc., on a printed circuit board of the Device could be captured in the Device description. A Device description should contain an identifier or name, and the location of deployment, either expressed in global coordinates or local human readable text (e.g. Auditorium).

It is important to observe that for several of these pieces of information that serve as attributes or properties of the different classes of the IoT Information Model, semantic data models or ontologies could be used in a real system implementation. For example, sensor values as Attributes could be annotated with metadata that point to the NASA SWEET ontology as already mentioned above. Location information could comply with an ontology such as GeoNames ontology (GeoNames 2013), and Device description could refer to specific Device ontologies.

7.3.3 Functional model

The IoT Functional Model aims at describing mainly the Functional Groups (FG) and their interaction with the ARM, while the Functional View of a Reference Architecture describes the functional components of an FG, interfaces, and interactions between the components. The Functional View is typically derived from the Functional Model in conjunction with high-level requirements. Interested readers in the requirement collection and generation process can refer to Pastor et al. (2011), and in the specific Unified Requirements (Figure 7.3) collected by the IoT-A can refer to IoT-A (2013).

This section briefly describes the most important functional groups, while the Reference Architecture chapter will elaborate on the composition of each FG. The IoT-A Functional Model is shown in Figure 7.11.

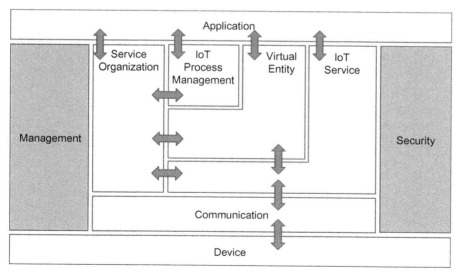

FIGURE 7.11

IoT-A Functional Model.

(Carrez et al. 2013)

The Application, Virtual Entity, IoT Service, and Device FGs are generated by starting from the User, Virtual Entity, Resource, Service, and Device classes from the IoT Domain Model. The need for communicating Devices and digital artifacts was the motivation for the Communication FG. The need to compose simple IoT services in order to create more complex ones, as well as the need to integrate IoT services (simple or complex) with existing Information and Communications Technology (ICT) infrastructure, is the main driver behind the introduction of the Service Organization and IoT Process Management FGs respectively. All the above-mentioned FGs need to be supported by management and security functionality captured by the corresponding FGs. The figure shows the flow of information between FGs apart from the cases of the Management and Security FGs that have information flowing from/to all other FGs, but these flows are omitted for clarity purposes.

7.3.3.1 Device functional group

The Device FG contains all the possible functionality hosted by the physical Devices that are used for instrumenting the Physical Entities. This Device functionality includes sensing, actuation, processing, storage, and identification components, the sophistication of which depends on the Device capabilities.

7.3.3.2 Communication functional group

The Communication FG abstracts all the possible communication mechanisms used by the relevant Devices in an actual system in order to transfer information to the digital world components or other Devices. Examples of such functions include wired bus or wireless mesh technologies through which sensor Devices are connected to Internet Gateway Devices. Communication technologies used between Applications and other functions such as functions from the IoT Service FG are out of scope because they are the typical Internet technologies. The reader is encouraged to refer to the corresponding sections in Chapter 5 related to Devices and local and wide area networking technologies.

7.3.3.3 IoT Service functional group

The IoT Service FG corresponds mainly to the Service class from the IoT Domain Model, and contains single IoT Services exposed by Resources hosted on Devices or in the Network (e.g. processing or storage Resources). Support functions such as directory services, which allow discovery of Services and resolution to Resources, are also part of this FG.

7.3.3.4 *Virtual Entity functional group*

The Virtual Entity FG corresponds to the Virtual Entity class in the IoT Domain Model, and contains the necessary functionality to manage associations between Virtual Entities with themselves as well as associations between Virtual Entities and related IoT Services, i.e. the Association objects for the IoT Information Model. Associations between Virtual Entities can be static or dynamic depending on the mobility of the Physical Entities related to the corresponding Virtual Entities. An example of a static association between Virtual Entities is the hierarchical inclusion relationship of a building, floor, room/corridor/open space, i.e. a building contains multiple floors that contain rooms, corridors, and open spaces. An example of a dynamic association between Virtual Entities is a car moving from one block of a city to another (the car is one Virtual Entity while the city block is another). A major difference between IoT Services and Virtual Entity Services is the semantics of the requests and responses to/from these services. Referring back to the parking lot example, the Parking Sensor Service provides as a response only a number "0" or "1" given the identifier of a Loop Sensor (e.g. #11). The Virtual Entity Parking Spot #01 responds to a request about its occupancy status as "free." The IoT Service provides data or information associated to specific Devices or Resources, including limited semantic information (e.g. Parking sensor #11, value = "0", units = none); the Virtual IoT Service provides information with richer semantics ("Parking spot #01 is free"), and is closer to being human-readable and understandable.

7.3.3.5 *IoT Service Organization functional group*

The purpose of the IoT Service Organization FG is to host all functional components that support the composition and orchestration of IoT and Virtual Entity services. Moreover, this FG acts as a service hub between several other functional groups such as the IoT Process Management FG when, for example, service requests from Applications or the IoT Process Management are directed to the Resources implementing the necessary Services. Therefore, the Service Organization FG supports the association of Virtual Entities with the related IoT Services, and contains functions for discovery, composition, and choreography of services. Simple IoT or Virtual Entity Services can be composed to create more complex services, e.g. a control loop with one Sensor Service and one Actuator service with the objective to control the temperature in a building. Choreography is the brokerage of Services so that Services can subscribe to other services in a system.

7.3.3.6 IoT Process Management functional group

The IoT Process Management FG is a collection of functionalities that allows smooth integration of IoT-related services (IoT Services, Virtual Entity Services, Composed Services) with the Enterprise (Business) Processes.

7.3.3.7 Management functional group

The Management FG includes the necessary functions for enabling fault and performance monitoring of the system, configuration for enabling the system to be flexible to changing User demands, and accounting for enabling subsequent billing for the usage of the system. Support functions such as management of ownership, administrative domain, rules and rights of functional components, and information stores are also included in the Management FG.

7.3.3.8 Security functional group

The Security FG contains the functional components that ensure the secure operation of the system as well as the management of privacy. The Security FG contains components for Authentication of Users (Applications, Humans), Authorization of access to Services by Users, secure communication (ensuring integrity and confidentiality of messages) between entities of the system such as Devices, Services, Applications, and last but not least, assurance of privacy of sensitive information relating to Human Users. These include privacy mechanisms such as anonymization of collected data, anonymization of resource and Service accesses (Services cannot deduce which Human User accessed the data), and un-linkability (an outside observer cannot deduce the Human User of a service by observing multiple service requests by the same User).

7.3.3.9 Application functional group

The Application FG is just a placeholder that represents all the needed logic for creating an IoT application. The applications typically contain custom logic tailored to a specific domain such as a Smart Grid. An application can also be a part of a bigger ICT system that employs IoT services such as a supply chain system that uses RFID readers to track the movement of goods within a factory in order to update the Enterprise Resource Planning (ERP) system.

7.3.3.10 Modular IoT functions

It is important to note that not all the FGs are needed for a complete actual IoT system. The Functional Model, as well as the Functional View of the Reference Architecture, contains a complete map of the potential functionalities for a system realization. The functionalities that will eventually be used in an actual system are dependent on the actual system requirements. What is important to observe is that the FGs are organized in such a way that more complex functionalities can be built based on simpler ones, thus making the model modular. This is shown already in Figure 7.11, where all the bidirectional arrows show the information flow between FGs, and it is illustrated further in Figure 7.12.

The bare minimum functionalities are Device, Communication, IoT Services, Management, and Security (Figure 7.12a). With these functionalities, an actual system can provide access to sensors, actuators and tag services for an application or backend system of a larger Enterprise. The application or larger system parts have to build the Virtual Entity functions for capturing the information about the Virtual Entities or the "Things" in the IoT architecture. Often the Virtual Entity concept is not captured in the application or a larger system with a dedicated FG, but functions for handling Virtual Entities are embedded in the application or larger system logic; therefore, in Figures 7.12a—c, the Virtual Entity is represented with dashed lines. For example, the deployment of a heating and cooling system is captured in a paper document out of which the larger system developers hardcode logic such as, "if sensor A value is above 25°C, turn on air

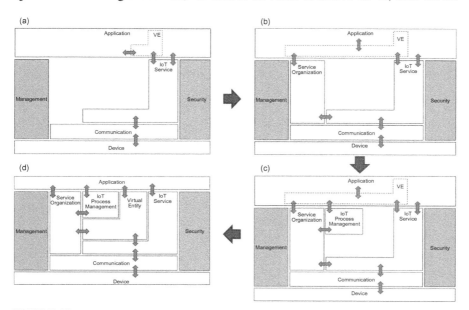

FIGURE 7.12

Building progressively complex IoT Systems.

conditioner B." The Virtual Entity in this case is a room that the developer does not care to capture. The next step in complexity of an IoT system is the addition of composed services based on simpler services. This is shown in (Figure 7.12b), where the Service Organization FG is added. Composed services can be used to abstract simpler services, e.g. filtering of events coming from multiple IoT Services. The following step in increasing complexity is the addition of the Business Processes functionality (Figure 7.12c) that enables Enterprise functionality to be present very close to the real IoT system, thus enabling local business control loops.

7.3.4 Communication model

The communication model for an IoT Reference Model consists of the identification of the endpoints of interactions, traffic patterns (e.g. unicast vs. multicast), and general properties of the underlying technologies used for enabling such interactions. Chapter 5 and parts of chapter 6 (e.g. ETSI M2M, IETF architectures) describe in detail architectures, traffic patterns, and specific networking technologies used for connecting the different endpoints together. As a result, this section only focuses on the identification of the endpoints of the communication paths.

The potential communicating endpoints or entities are the Users, Resources, and Devices from the IoT Domain Model. Users include Human Users and Active Digital Artifacts (Services, internal system components, external applications). Devices with a Human—Machine Interface mediate the interactions between a Human User and the physical world (e.g. keyboards, mice, pens, touch screens, buttons, microphones, cameras, eye tracking, and brain wave interfaces, etc.), and therefore the Human User is not a communication model endpoint. The User (Active Digital Artifact, Service)-to-Service interactions include the User-to-Service and Service-to-Service interactions (in the case of an enterprise service/ application accessing another service, or in the case of IoT service composition) as well as the Service—Resource—Device interactions. The User-to-Service and Service-to-Service communication is typically based on Internet protocols today as described in Chapter 5, apart from the case of Service-to-Service interactions when one or both Services are hosted on constrained/low-end Devices such as embedded systems. Typically, constrained Devices have a different communication stack from the ones used for the Internet-type of networks. Examples of such constrained network technologies appear in Chapter 5. Therefore, the communication model for these interactions includes several types of gateways (e.g. network, application layer gateways) to bridge between two or more disparate communication technologies. A similar problem occurs between the Service-to-Resource communications. The Devices may be so constrained that they cannot host the Services, while the Resources could be hosted or not depending on the Device capabilities. This inability of the Device to host Resources or Services results in moving the corresponding Resources and/or Services out of the Device and into more powerful Devices or machines in the cloud. Then the Resource-to-Device or the Service-to-Resource communication needs to involve multiple types of communication stacks.

7.3.5 Safety, privacy, trust, security model

An IoT system enables interactions between Human Users and Active Digital Artifacts (Machine Users) with the physical environment. The fact that Human Users are part of the system that could potentially harm humans if malfunctioning, or expose private information, motivates the Safety and Privacy needs for the IoT Reference Model and Architecture. The Trust and Security Model are needed in every ICT system with the objective to protect the digital world.

7.3.5.1 Safety

System safety is highly application- or application domain-specific, and is typically closely related to an IoT system that includes actuators that could potentially harm animate objects (humans, animals). For example, the operation of an IoT system controlling an elevator could harm humans if it allowed the elevator doors to open with a normal user interaction when the elevator room is not behind the doors. Critical infrastructure protection is also related to safety because the loss of such infrastructure due to a malicious user attack could be detrimental to humans, e.g. attacks to a Smart Grid could result in damages ranging from simple loss of electricity in a home to electricity loss in a hospital. By not being application-specific, the IoT Reference Model can only provide IoT-related guidelines for ensuring a safe system to the extent possible and controllable by a system designer. A system designer of such critical systems typically follows an iterative process with two steps: (a) identification of hazards followed by, (b) the mitigation plan. This process is very similar to the threat model and mitigation plan that a security designer performs for an ICT system. Not all hazards or mitigation steps include IoT technology, but a system designer could add safety assertions in relevant points in the interaction between Users, Services, Resources, and Devices. For example, a Human User interaction of pressing the elevator button should result in the elevator door opening only when a Sensor Device detects the elevator room to be behind the doors. If the system designer would like to provide for a safer elevator system, the system should include mechanical safety locks that work even in the absence of the loss of electricity. However, these additional measures don't depend on an IoT system described in this book.

7.3.5.2 Privacy

Because interactions with the physical world may often include humans, protecting the User privacy is of utmost importance for an IoT system. The IoT-A Privacy Model (Carrez et al. 2013, Gruschka & Gessner 2012) depends on the following functional components: Identity Management, Authentication, Authorization, and Trust & Reputation. Identity Management offers the derivation of several identities of different types for the same architectural entity with the objective to protect the original User identity for anonymization purposes. Authentication is a function that allows the verification of the identity of a User whether this is the original or some derived identity. Authorization is the function that asserts and enforces access rights when Users (Services, Human Users) interact with Services, Resources, and Devices. The Trust and Reputation functional component maintain the static or dynamic trust relationships between interacting entities. These relationships can impact the behavior of inter-acting entities, for example, if a Device is deemed untrusted (e.g. when its Sensor on-Device Service reports out of range measurements), another entity (e.g. another Sensor Device or Gateway) could contain logic to reject sensor measurements from the particular Device. The level of Trust and Reputation typically reflects the level of expected behavior of an entity. In ICT systems, trust and reputation are typically represented by a trust/reputation score used for ranking similar entities (e.g. Devices providing similar sensor measurements).

7.3.5.3 Trust

According to the Internet Engineering Task Force (IETF) Internet Security Glossary (Shirey 2007), "Generally, an entity is said to 'trust' a second entity when the first entity makes the assumption that the second entity will behave exactly as the first entity expects." This definition includes an "expectation" which is difficult to capture in a technical context. Nevertheless, Gruschka & Gessner (2012) suggest that in a technical context, Trust and Reputation could be represented by a score, as seen earlier. This score could be used to impact the behavior of technical components interacting with each other. A trust model is often coupled with the notion of trust in an ICT system, and represents the model of dependencies and expectations of interacting entities (Baugé et al. 2010). The necessary aspects of a trust model according to IoT-A (Gruschka & Gessner 2012) are as follows:

- **Trust Model Domains**: Because ICT and IoT systems may include a large number of interacting entities with different properties, maintaining trust relationships for every pair of interacting entities may be prohibitive. Therefore, groups of entities with similar trust properties can define different trust domains.
- **Trust Evaluation Mechanisms**: These are well-defined mechanisms that describe how a trust score could be computed for a specific entity. The evaluation mechanism needs to take into account the source of information used for computing the trust level/score of an entity; two related aspects are the federated trust and trust anchor. A related concept is the IoT support for evaluation of the trust level of a Device, Resource, and Service.
- **Trust Behavior Policies**: These are policies that govern the behavior between interacting entities based on the trust level of these interacting entities; for example, how a User could use sensor measurements retrieved by a Sensor Service with a low trust level.
- **Trust Anchor**: This is an entity trusted by default by all other entities belonging to the same trust model, and is typically used for the evaluation of the trust level of a third entity.
- **Federation of Trust**: A federation between two or more Trust Models includes a set of rules that specify the handling of trust relationships between entities with different Trust Models. Federation becomes important in large-scale systems.

7.3.5.4 Security

The Security Model for IoT consists of communication security that focuses mostly on the confidentiality and integrity protection of interacting entities and functional components such as Identity Management, Authentication, Authorization, and Trust & Reputation, as seen earlier.

This chapter provided an overview of the IoT ARM. Chapter 8 provides the IoT Reference Architecture.

IoT Reference Architecture

8

CHAPTER OUTLINE

8.1 Introduction

In this chapter we describe the IoT Reference Architecture. As mentioned earlier, the Reference Architecture is a starting point for generating concrete architectures and actual systems. A concrete architecture addresses the concerns of multiple stakeholders of the actual system, and it is typically presented as a series of views that address different stakeholder concerns (Kruchten 1995; SEI CMU 2013; Rozanski & Woods 2005, 2011). A Reference Architecture, on the other hand, serves as a guide for one or more concrete system architects. However, the concept of views for the presentation of an architecture is also useful for the IoT

From Machine-to-Machine to the Internet of Things: Introduction to a New Age of Intelligence.
DOI: http://dx.doi.org/10.1016/B978-0-12-407684-6.00008-5

Reference Architecture. Views are useful for reducing the complexity of the Reference Architecture blueprints by addressing groups of concerns one group at a time. However, since the IoT Reference Architecture does not contain details about the environment where the actual system is deployed, some views cannot be presented in detail or at all; for example, the view that shows the concrete Physical Entities and Devices for a specific scenario.

The stakeholders for a concrete IoT system are the people who use the system (Human Users); the people who design, build, and test the Resources, Services, Active Digital Artifacts, and Applications; the people who deploy Devices and attach them to Physical Entities; the people who integrate IoT capabilities of functions with an existing ICT system (e.g. of an enterprise); the people who operate, maintain, and troubleshoot the Physical and Virtual Infrastructure; and the people who buy and own an IoT system or parts thereof (e.g. city authorities).

In order to address the concerns of mainly the concrete IoT architect, and secondly the concerns of most of the above stakeholders, we have chosen to present the Reference Architecture as a set of architectural views (Kruchten 1995; SEI CMU 2013; Rozanski & Woods 2005, 2011):

- **Functional View**: Description of what the system does, and its main functions.
- **Information View**: Description of the data and information that the system handles.
- **Deployment and Operational View**: Description of the main real world components of the system such as devices, network routers, servers, etc.

The approach for this chapter is to describe the different views from a generic to a more specific view. The most specific architectural views appear later in the book in the specific use case chapters.

8.2 Functional view

The functional view for the IoT Reference Architecture is presented in Figure 8.1, and is adapted from IoT-A (Carrez et al. 2013). It consists of the Functional Groups (FGs) presented earlier in the IoT Functional Model, each of which includes a set of Functional Components (FCs). It is important to note that not all the FCs are used in a concrete IoT architecture, and therefore the actual system as explained earlier.

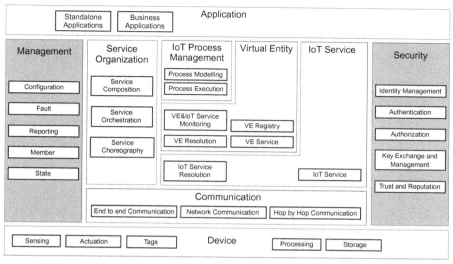

FIGURE 8.1

IoT Functional View.

(Adapted from IoT-A; Carrez et al. 2013)

8.2.1 Device and Application functional group

The Device and Application FGs are already covered in the IoT Functional Model. For convenience the Device FG contains the Sensing, Actuation, Tag, Processing, Storage FCs, or simply components. These components represent the resources of the device attached to the Physical Entities of interest. The Application FG contains either standalone applications (e.g. for iOS, Android, Windows phone), or Business Applications that connect the IoT system to an Enterprise system.

8.2.2 Communication functional group

The Communication FG contains the End-to-End Communication, Network Communication, and Hop-by-Hop communication components:

- The Hop-by-Hop Communication is applicable in the case that devices are equipped with mesh radio networking technologies such as IEEE 802.15.4 for which messages have to traverse the mesh from node-to-node (hop-by-hop) until they reach a gateway node which forwards the message (if needed) further to the Internet.

The hop-by-hop FC is responsible for transmission and reception of physical and MAC layer frames to/from other devices. This FC has two main interfaces: (a) one "southbound" to/from the actual radio on the device, and (b) one "northbound" to/from the Network FC in the Communication FG.

- The Network FC is responsible for message routing & forwarding and the necessary translations of various identifiers and addresses. The translations can be (a) between network layer identifiers to MAC and/or physical network identifiers, (b) between high-level human readable host/node identifiers to network layer addresses (e.g. Fully Qualified Domain Names (FQDN) to IP addresses, a function implemented by a Domain Name System (DNS) server), and (c) translation between node/service identifiers and network locators in case the higher layers above the networking layer use node or service identifiers that are decoupled from the node addresses in the network (e.g. Host Identity Protocol (HIP; Moskovitz & Nikander 2006) identifiers and IP addresses). Potential fragmentation and reassembly of messages due to limitations of the underlying layers is also handled by the Network FC. Finally, the Network FC is responsible for handling messages that cross different networking or MAC/PHY layer technologies, a function that is typically implemented on a network gateway type of device. An example is an IPv4 to IPv6 translation hosted in a gateway with two network interfaces, one supporting IPv4 and one supporting IPv6/6LoWPAN/IEEE 802.15.4. The Network FC interfaces the End-to-End Communication FC on the "northbound" direction, and the Hop-by-Hop Communication FC on the "southbound" direction.
- The End-to-End Communication FC is responsible for end-to-end transport of application layer messages through diverse network and MAC/PHY layers. In turn, this means that it may be responsible for end-to-end retransmissions of missing frames depending on the configuration of the FC. For example, if the End-to-End Communication FC is mapped in an actual system to a component implementing the Transmission Control Protocol (TCP) protocol, reliable transfer of frames dictates the retransmission of missing frames. Finally, this FC is responsible for hosting any necessary proxy/cache and any protocol translation between networks with different transport/application layer technologies. An example of such functionality is the HTTP-CoAP proxy, which performs transport-layer protocol translation. The End-to-End FC interfaces the Network FC on the "southbound" direction.

8.2.3 **IoT Service functional group**

The IoT Service FG consists of two FCs: The IoT Service FC and the IoT Service Resolution FC:

- The IoT Service FC is a collection of service implementations, which interface the related and associated Resources. For a Sensor type of a Resource, the IoT Service FC includes Services that receive requests from a User and returns the Sensor Resource value in synchronous or asynchronous (e.g. subscription/notification) fashion. The services corresponding to Actuator Resources receive User requests for actuation, control the Actuator Resource, and may return the status of the Actuator after the action. A Tag IoT Service can behave both as a Sensor (for reading the identifier of the Tag), or as an Actuator (for writing a new identifier or information on the Tag, if possible). As mentioned earlier, Resources can also perform processing and storage (Processing or Storage Resources), and therefore their corresponding Service exposes the corresponding interfaces, for example, interfaces to provide input data and retrieve output data from a Complex Event Processing (CEP) Resource. An IoT Service for a particular Resource could also expose as a service the historical values of sensor values or actuator commands or tag identifiers.
- The IoT Service Resolution FC contains the necessary functions to realize a directory of IoT Services that allows dynamic management of IoT Service descriptions and discovery/lookup/resolution of IoT Services by other Active Digital Artifacts. The Service descriptions of IoT Services contain a number of attributes as seen earlier in the IoT Functional Model section. Dynamic management includes methods such as creation/update/deletion (CUD) of Service description, and can be invoked by both the IoT Services themselves, or functions from the Management FG (e.g. bulk creation of IoT Service descriptions upon system start-up). The discovery/lookup and resolution functions allow other Services or Active Digital Artifacts to locate IoT Services by providing different types of information to the IoT Service Resolution FC. By providing the Service identifier (attribute of the Service description) a lookup method invocation to the IoT Service Resolution returns the Service description, while the resolution method invocation returns the contact information (attribute of the service description) of a service for direct Service invocation (e.g. URL). The discovery method, on the other hand, assumes that the Service identifier is unknown, and the discovery request contains a set of desirable Service description attributes that matching Service descriptions should contain.

8.2.4 **Virtual Entity functional group**

The Virtual Entity FG contains functions that support the interactions between Users and Physical Things through Virtual Entity services. An example of such an interaction is the query to an IoT system of the form, "What is the temperature in the conference room Titan?" The Virtual Entity is the conference room "Titan," and the conference room attribute of interest is "temperature." Assuming that the room is actually instrumented with a temperature sensor, if the User had the knowledge of which temperature sensor is installed in the room (e.g. TempSensor #23), then the User could re-formulate and re-target this query to, "What is the value of TempSensor #23?" dispatched to the relevant IoT Service representing the temperature resource on the TempSensor #23. The Virtual Entity interaction paradigm requires functionality such as discovery of IoT Services based on Virtual Entity descriptions, managing the Virtual Entity-IoT Service associations, and processing Virtual Entity-based queries. The following FCs are defined for realizing these functionalities:

- The Virtual Entity Service FC enables the interaction between Users and Virtual Entities by means of reading and writing the Virtual Entity attributes (simple or complex), which can be read or written, of course. Some attributes (e.g. the GPS coordinates of a room) are static and non-writable by nature, and some other attributes are non-writable by access control rules. In general attributes that are associated with IoT Services, which in turn represent Sensor Resources, can only be read. There can be, of course, special Virtual Entities associated with the same Sensor Resource through another IoT Service that allow write operations. An example of such a special case is when the Virtual Entity represents the Sensor device itself (for management purposes). In general, attributes that are associated with IoT Services, which in turn represent Actuator Resources, can be read and written. A read operation returns the actuator status, while a write operation results in a command sent to the actuator. Virtual Entity attributes corresponding to Tags can be read in most of the cases by Users, and can be written by special cases by other types of Users (e.g. Management applications), if possible of course, as is the case of re-writable RFID tags. Apart from the function to operate on the Virtual Entity attributes, a Virtual Entity Service can also expose the historical variations of the attributes of a Virtual Entity.
- The Virtual Entity Registry FC maintains the Virtual Entities of interest for the specific IoT system and their associations. The component offers services such as creating/reading/updating/deleting Virtual Entity

descriptions and associations. Certain associations can be static; for example, the entity "Room #123" is contained in the entity "Floor #7" by construction, while other associations are dynamic, e.g. entity "Dog" and entity "Living Room" due to at least Entity mobility. Update and Deletion operations take the Virtual Entity identifier as a parameter.

- The Virtual Entity Resolution FC maintains the associations between Virtual Entities and IoT Services, and offers services such as creating/reading/updating/deleting associations as well as lookup and discovery of associations. The Virtual Entity Resolution FC also provides notification to Users about the status of the dynamic associations between a Virtual Entity and an IoT Service, and finally allows the discovery of IoT Services provided the certain Virtual Entity attributes.
- The Virtual Entity and IoT Service Monitoring FC includes: (a) functionality to assert static Virtual Entity−IoT Service associations, (b) functionality to discover new associations based on existing associations or Virtual Entity attributes such as location or proximity, and (c) continuous monitoring of the dynamic associations between Virtual Entities and IoT Services and updates of their status in case existing associations are not valid any more.

The difference between IoT Service & Resource associations and Virtual Entity & IoT-Service associations is that the former are typically static and created upon the creation of the IoT Service instantiation, while the latter are generally dynamic (without excluding static associations, of course) because of potential Virtual Entity mobility. The result of this difference is that in the IoT Service FG, there is no FC to discover or monitor new IoT-Service-to-Resource associations, while in the Virtual Entity FG there is a corresponding one.

8.2.5 IoT process management functional group

The IoT Process Management FG aims at supporting the integration of business processes with IoT-related services. It consists of two FCs:

- The Process Modeling FC provides that right tools for modeling a business process that utilizes IoT-related services.
- The Process Execution FC contains the execution environment of the process models created by the Process Modelling FC and executes the created processes by utilizing the Service Organization FG in order to resolve high-level application requirements to specific IoT services.

It is important to note the IoT services mentioned above are not only the services from the IoT Service FG, but also from the Virtual Entity FG and the Service Organization FG.

8.2.6 Service Organization functional group

The Service Organization FG acts as a coordinator between different Services offered by the system. It consists of the following FCs:

- The Service Composition FC manages the descriptions and execution environment of complex services consisting of simpler dependent services. An example of a complex composed service is a service offering the average of the values coming from a number of simple Sensor Services. The complex composed service descriptions can be well-specified or dynamic/flexible depending on whether the constituent services are well-defined and known at the execution time or discovered on-demand. The objective of a dynamic composed service can be the maximization of the quality of information achieved by the composition of simpler Services, as is the case with the example "average" service described earlier.
- The Service Orchestration FC resolves the requests coming from IoT Process Execution FC or User into the concrete IoT services that fulfill the requirements.
- The Service Choreography FC is a broker for facilitating communication among Services using the Publish/Subscribe pattern. Users and Services interested in specific IoT-related services subscribe to the Choreography FC, providing the desirable service attributes even if the desired services do not exist. The Choreography FC notifies the Users when services fulfilling the subscription criteria are found.

It is important to note that the IoT services mentioned above are not only the services from the IoT Service FG, but also from the Virtual Entity FG and the Service Composition FC.

8.2.7 Security functional group

The Security FG contains the necessary functions for ensuring the security and privacy of an IoT system. It consists of the following FCs:

- The Identity Management FC manages the different identities of the involved Services or Users in an IoT system in order to achieve anonymity by the use of multiple pseudonyms. The component maintains a hierarchy of identities (an identity pool), as well as group identities (Gruschka & Gessner 2012).

- The Authentication FC verifies the identity of a User and creates an assertion upon successful verification. It also verifies the validity of a given assertion.
- The Authorization FC manages and enforces access control policies. It provides services to manage policies (CUD), as well as taking decisions and enforcing them regarding access rights of restricted resources. The term "resource" here is used as a representation of any item in an IoT system that needs a restricted access. Such an item can be a database entry (Passive Digital Artifact), a Service interface, a Virtual Entity attribute (simple or complex), a Resource/Service/Virtual Entity description, etc.
- The Key Exchange & Management is used for setting up the necessary security keys between two communicating entities in an IoT system. This involves a secure key distribution function between communicating entities.
- The Trust & Reputation FC manages reputation scores of different interacting entities in an IoT system and calculates the service trust levels. A more detailed description of this FC is contained in the Safety, Privacy, Trust, Security Model presented in Chapter 7.

8.2.8 Management functional group

The Management FG contains system-wide management functions that may use individual FC management interfaces. It is not responsible for the management of each component, rather for the management of the system as a whole. It consists of the following FCs:

- The Configuration FC maintains the configuration of the FCs and the Devices in an IoT system (a subset of the ones included in the Functional View). The component collects the current configuration of all the FCs and devices, stores it in a historical database, and compares current and historical configurations. The component can also set the system-wide configuration (e.g. upon initialization), which in turn translates to configuration changes to individual FCs and devices.
- The Fault FC detects, logs, isolates, and corrects system-wide faults if possible. This means that individual component fault reporting triggers fault diagnosis and fault recovery procedures in the Fault FC.
- The Member FC manages membership information about the relevant entities in an IoT system. Example relevant entities are the FGs, FCs,

Services, Resources, Devices, Users, and Applications. Membership information is typically stored in a database along with other useful information such as capabilities, ownership, and access rules & rights, which are used by the Identity Management and Authorization FCs.

- The State FC is similar to the Configuration FC, and collects and logs state information from the current FCs, which can be used for fault diagnosis, performance analysis and prediction, as well as billing purposes. This component can also set the state of the other FCs based on system-wise state information.
- The Reporting FC is responsible for producing compressed reports about the system state based on input from FCs.

8.3 Information view

The information view consists of (a) the description of the information handled in the IoT System, and (b) the way this information is handled in the system; in other words, the information lifecycle and flow (how information is created, processed, and deleted), and the information handling components. Because the information handled by an IoT system is captured mainly by the IoT Information Model described in Chapter 7 as part of the IoT Reference Model, we only provide a synopsis of the specific information pieces without going into details. As a second part, we describe the way some of the above-mentioned pieces of information are handled in an IoT system.

8.3.1 Information description

The pieces of information handled by an IoT system complying to an ARM such as the IoT-A (Carrez et al. 2013) are the following:

- Virtual Entity context information, i.e. the attributes (simple or complex) as represented by parts of the IoT Information model (attributes that have values and metadata such as the temperature of a room). This is one of the most important pieces of information that should be captured by an IoT system, and represents the properties of the associated Physical Entities or Things.
- IoT Service output itself is another important part of information generated by an IoT system. For example, this is the information generated by interrogating a Sensor or a Tag Service.
- Virtual Entity descriptions in general, which contain not only the attributes coming from IoT Devices (e.g. ownership information).

- Associations between Virtual Entities and related IoT Services.
- Virtual Entity Associations with other Virtual Entities (e.g. Room #123 is on Floor #7).
- IoT Service Descriptions, which contain associated Resources, interface descriptions, etc.
- Resource Descriptions, which contain the type of resource (e.g. sensor), identity, associated Services, and Devices.
- Device Descriptions such as device capabilities (e.g. sensors, radios).
- Descriptions of Composed Services, which contain the model of how a complex service is composed of simpler services.
- IoT Business Process Model, which describes the steps of a business process utilizing other IoT-related services (IoT, Virtual Entity, Composed Services).
- Security information such as keys, identity pools, policies, trust models, reputation scores, etc.
- Management information such as state information from operational FCs used for fault/performance purposes, configuration snapshots, reports, membership information, etc.

8.3.2 Information flow and lifecycle

On a high level, the flow of information in an IoT system follows two main directions. From devices that produce information such as sensors and tags, information follows a context-enrichment process until it reaches the consumer application or part of the larger system, and from the application or part of a larger system information it follows a context-reduction process until it reaches the consumer types of devices (e.g. actuators). The enrichment process is shown in Figure 8.2. Devices equipped with sensors transform changes in the physical properties of the Physical Entities of Interest into electrical signals. These electrical signals are transformed in one or multiple values (Figure 8.2a) on the device level. These values are then enriched with metadata information such as units of measurement, timestamp, and possibly location information (Figure 8.2b). These enriched values are offered by a software component (Resource) either on the device or the network. The Resource exposes certain IoT Services to formalize access to this enriched information (Figure 8.2c). At this point, the information is annotated with simple attributes such as location and time, and often this type of metadata is sufficient for certain IoT applications or for the use in certain larger systems. This enriched information becomes context information as soon as it is further associated with certain Physical Entities in the form of Virtual Entity attributes (simple or complex, static or dynamic).

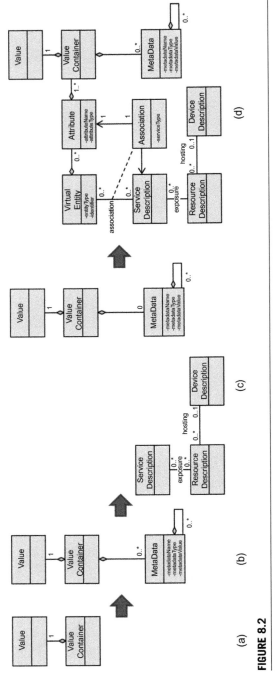

FIGURE 8.2

Information-enrichment process.

Further support information such as Associations between certain attributes and IoT Services further enriches the context information of the Virtual Entity (Figure 8.2d).

Further enrichment occurs in applications or larger systems that employ, for example, data analytics, machine learning, and knowledge management, which produces actionable information. Parts of the context and actionable information may be stored to an information store for future use. Actionable information flows into business processes that implement an action plan. Action plans push context information about Virtual Entities to associated IoT Services, to corresponding Actuation Resources, and finally to the real actuators that perform the changes in the physical world (context information reduction flow). Actual IoT systems employ a different degree of enrichment, reduction, or storage. Certain IoT systems only employ enrichment while leaving the action to humans, others employ only context reduction (e.g. remote control of a heating element), or others employ the full feedback loop.

Virtual Entity context information is typically generated by data-producing devices such as sensor devices and consumed either by data-consumption devices such as actuators or services (IoT or other types of services such as machine learning processing services). Raw or enriched information and/or actionable information may be stored in caches or historical databases for later usage or processing, traceability, or accounting purposes. The historical/cache database information lifetime is often application- or regulation-specific. Typically the information maintained in a cache is ephemeral, while the information stored in historical databases can last for longer but highly application-specific durations. Certain raw sensor readings may be destroyed after fulfilling a User request, and other sensor readings may be stored for 5 years (for example) according to a data-retention policy dictated by regulation. Similar rules may apply for the operation-specific information such as Device, Resource, Service Descriptions, etc. These contain the necessary information for an IoT system to operate, but not the information that human users are typically interested in. Nevertheless, this kind of information is created by FCs, are typically stored for the purpose of fault management, and are typically automatically destroyed, for example, when a soft state handling policy is applicable. In this context soft state information means that the subsystem that manages this kind of information destroys old information according to time of creation and a retention policy (e.g. Service Descriptions older than 1 day are destroyed from the IoT Service Resolution), while the Users (FCs, management applications) that create this kind of information are responsible for refreshing it periodically in order to avoid automatic destruction.

8.3.3 **Information handling**

An IoT system is typically deployed to monitor and control Physical Entities. Monitoring and controlling Physical Entities is in turn performed by mainly the Devices, Communication, IoT Services, and Virtual Entity FGs in the functional view. Certain FCs of these FGs, as well as the rest of the FGs (Service Organization, IoT Process Management, Management, Security FGs), play a supporting role for the main FGs in the Reference Architecture, and therefore in the flow of information. Moreover, an IoT system is typically one part of a larger system encompassing several other functions, such as Complex Event Processing (CEP), Data Collection and Processing, and Data Analytics and Knowledge Management, as seen earlier in the book in Chapter 5. Therefore, information handling of an IoT system depends largely on the specific problem at hand. From a Reference Architecture point of view, we can only present part of the information flow space concerning the IoT Reference Model, while the technology chapter (Chapter 5) provides details on the individual and more complex information handling components and interactions.

The presentation of information handling in an IoT system assumes that FCs exchange and process information. The exchange of information between FCs follows the interaction patterns below (Carrez et al. 2013; Figure 8.3):

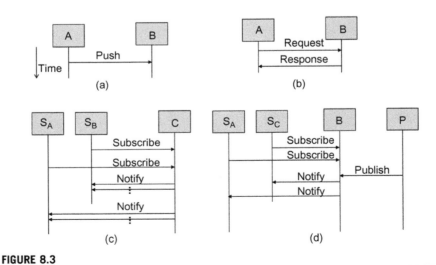

FIGURE 8.3

Information exchange patterns.

(Adapted from IoT-A; Carrez et al. 2013)

- **Push**: An FC A pushes the information to another FC B provided that the contact information of the component B is already configured in component A, and component B listens for such information pushes.
- **Request/Response**: An FC A sends a request to another FC B and receives a response from B after A serves the request. Typically the interaction is synchronous in the sense that A must wait for a response from B before proceeding to other tasks, but in practice this limitation can be realized with parts of component A waiting, and other parts performing other tasks. Component B may need to handle concurrent requests and responses from multiple components, which imposes certain requirements on the capabilities for the device or the network that hosts the FC.
- **Subscribe/Notify**: Multiple subscriber components (S_A, S_B) can subscribe for information to a component C, and C will notify the relevant subscribers when the requested information is ready. This is typically an asynchronous information request after which each subscriber can perform other tasks. Nevertheless, a subscriber needs to have some listening components for receiving the asynchronous response. The target component C also needs to maintain state information about which subscribers requested which information and their contact information. The Subscribe/Notify pattern is applicable when typically one component is the host of the information needed by multiple other components. Then the subscribers need only establish a Subscribe/Notify relationship with one component. If multiple components can be information producers or information hosts, the Publish/Subscribe pattern is a more scalable solution from the point of view of the subscribers.
- **Publish/Subscribe**: In the Publish/Subscribe (also known as a Pub/Sub pattern), there is a third component called the broker B, which mediates subscription and publications between subscribers (information consumers) and publishers (or information producers). Subscribers such as S_A and S_B subscribe to the broker about the information they are interested in by describing the different properties of the information. Publishers publish information and metadata to the broker, and the broker pushes the published information to (notification) the subscribers whose interests match the published information.

At this point we describe a few examples of information handling by the FCs. Please note that these interaction descriptions are not complete in the sense that they do not contain all the possible ways such interactions can take place, and they are not intended to be complete since these interactions are highly dependent on the actual IoT system requirements.

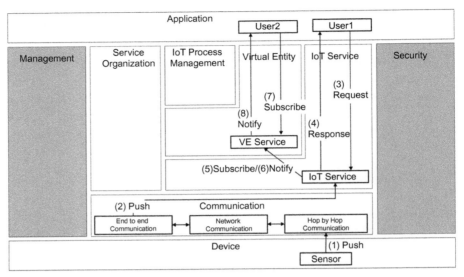

FIGURE 8.4

Device, IoT Service, and Virtual Entity Service Interactions.

In Figure 8.4 we assume that the generated sensed data is pushed by a sensor device (under Steps 1 and 2) that is part of a multi-hop mesh network such as IEEE 802.15.4 through the Hop-by-Hop, Network, and End-to-End communication FCs towards the Sensor Resource hosted in the network. Please note that the Sensor Resource is not shown in the figure, only the associated IoT Service. A cached version of the sensor reading on the Device is maintained on the IoT Service. When User1 (Step 3) requests the sensor reading value from the specific Sensor Device (assuming User1 provides the Sensor resource identifier), the IoT Service provides the cached copy of the sensor reading back to the User1 annotated with the appropriate metadata information about the sensor measurement, for example, timestamp of the last known reading of the sensor, units, and location of the Sensor Device. Also assume that that the Virtual Entity Service associated with the Physical Entity (e.g. a room in a building) where the specific Sensor Device has been deployed already contains the IoT Service as a provider of the "hasTemperature" attribute of its description. The Virtual Entity Service subscribes to the IoT Service for updates of the sensor readings pushed by the Sensor Device (Step 5). Every time the Sensor Device pushes sensor readings to the IoT Service, the IoT Service notifies (Step 6) the Virtual Entity Service, which updates the value of the attribute "hasTemperature" with the sensor reading of the Sensor Device. At a later stage, a User2 subscribing

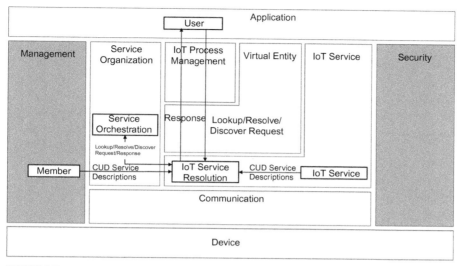

FIGURE 8.5

IoT Service Resolution.

(Step 7) to changes on the Virtual Entity attribute "hasTemperature" is notified every time the attribute changes value (Step 8). Please note that some of the information flow steps between the Virtual Entity and IoT Service are omitted in this figure for simplicity purposes. These interactions are shown in the figures below.

Figure 8.5 depicts the information flow when utilizing the IoT Service Resolution FC. The IoT Service Resolution implements two main interfaces, one for the CUD of Service Description objects in the IoT Service Resolution database/store, and one for lookup/resolution/discovery of IoT Services. As a reminder, the lookup and resolution operations provide the Service Description and the Service locator, respectively, given the Service identifier and the discovery operation returns a (set of) Service Description(s) given a list of desirable attributes that matching Service Descriptions should contain. The CUD operations can be performed by the IoT Service logic itself or by a management component (e.g. Member FC in Figure 8.5). The lookup/resolution and discovery operation can be performed by a User as a standalone query or the Service Orchestration as a part of a Composed Service or an IoT Process. If a discovery operation returns multiple matching Service Descriptions, it is upon the User or the Service Orchestration component to select the most appropriate IoT Service for the specific task. Although the interactions in Figure 8.5 follow the Request/Response patterns, the lookup/resolution/discovery operations can follow the Subscribe/Notify pattern in the sense that a User or the Service Orchestration FC subscribe to changes of existing IoT Services for lookup/resolution and for the discovery of new Service Descriptions in the case of a discovery operation.

Figure 8.6 describes the information flow when the Virtual Entity Service Resolution FC is utilized. The Virtual Entity Resolution FC allows the CUD of Virtual Entity Descriptions, and the lookup and discovery of Virtual Entity Descriptions. A lookup operation by a User or the Service Orchestration FC returns the Virtual Entity Description given the Virtual Entity identity, while the discovery operation returns the Virtual Entity Description(s) given a set of Virtual Entity attributes (simple or complex) that matching Virtual Entities should contain. Please note that the Virtual Entity Registry is also involved in the information flow because it is the storage component of Virtual Entity Descriptions, but it is omitted from the figure to avoid cluttering. The Virtual Entity Resolution FC mediates the requests/responses/subscriptions/notifications between Users and the Virtual Entity Registry, which has a simple create/read/update/delete (CRUD) interface given the Virtual Entity identity. The FCs that could perform CUD operations on the Virtual Entity Resolution FC are the IoT Services themselves due to internal configuration, the Member Management FC that maintains the associations as part of the system setup, and the Virtual Entity and IoT Service Monitoring component whose purpose is to discover dynamic associations between Virtual Entities and IoT Services. It is important to note that the

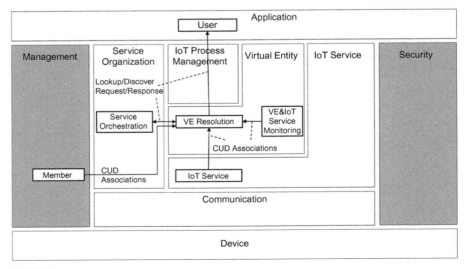

FIGURE 8.6

Virtual Entity Service Resolution.

Subscribe/Notify interaction patterns can also be applicable to the lookup/ discovery operations, the same as the Request/Response patterns provided the involved FCs implement Subscribe/Notify interfaces.

As a final example of the information flow, we show a Complex Event Processing (CEP) Resource mapped to an IoT Service C. The CEP Service needs the information from two IoT services, e.g. IoT Services corresponding to two Sensor Resources hosted on two Sensor Devices, and produces one output. The CEP IoT Service expects the inputs to be published/pushed to its interfaces, while the output interface conforms to a Subscribe/Notify interaction pattern. The individual IoT Services A and B expose interfaces that also comply with the Publish/Subscribe interaction pattern. The FC that can connect these three components is the Service Choreography FC that realizes a Publish/Subscribe interaction pattern. As a first step, the IoT Service C subscribes to the Service Choreography FC that it requires IoT Services A and B as inputs. In the meantime, a User subscribes to the Service Choreography FC that it needs the output of the CEP IoT Service C. When the individual IoT Services A and B publish their output to the Service Choreography FC, these outputs are published/ forwarded to the IoT Service C, which needs them to produce information of type C. After performing CEP filtering, the IoT Service C publishes the output of type C to the Service Choreography FC, which publishes/for-ward it to the User (Figure 8.7).

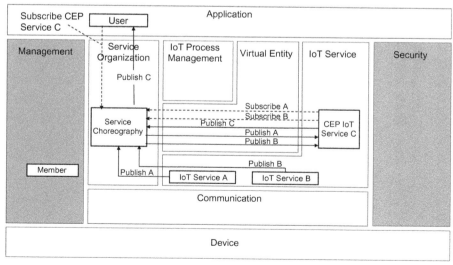

FIGURE 8.7

Service Choreography and Processing IoT Services.

(Adapted from Carrez et al. 2013)

8.4 Deployment and operational view

The Deployment and Operational View depends on the specific actual use case and requirements, and therefore we present here one way of realizing the Parking Lot example seen earlier. It is by no means an exhaustive or complete example. The use case chapters presented in Part III of this book provide real world deployment examples.

Figure 8.8 depicts the Devices view as Physical Entities deployed in the parking lot, as well as the occupancy sign. There are two sensor nodes (#1 and #2), each of which are connected to eight metal/car presence sensors. The two sensor nodes are connected to the payment station through wireless or wired communication. The payment station acts both as a user interface for the driver to pay and get a payment receipt as well as a communication gateway that connects the two sensor nodes and the payment interface physical devices (displays, credit card slots, coin/note input/output, etc.) with the Internet through Wide Area Network (WAN) technology. The occupancy sign also acts as a communication gateway for the actuator node (display of

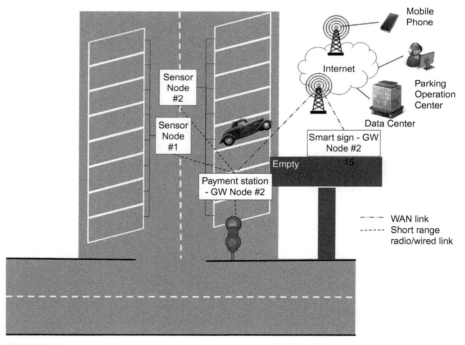

FIGURE 8.8

Parking Lot Deployment and Operational View, Devices.

free parking spots), and we assume that because of the deployment, a direct connection to the payment station is not feasible (e.g. wired connectivity is too prohibitive to be deployed or sensitive to vandalism). The physical gateway devices connect through a WAN technology to the Internet and towards a data center where the parking lot management system software is hosted as one of the virtual machines on a Platform as a Service (PaaS; Chapter 5) configuration. The two main applications connected to this management system are human user mobile phone applications and parking operation center applications. We assume that the parking operation center manages several other parking lots using similar physical and virtual infrastructure.

Figure 8.9 shows two views superimposed, the deployment and functional views, for the parking lot example. Please note that several FGs and FCs are omitted here for simplicity purposes, and certain non-IoT-specific

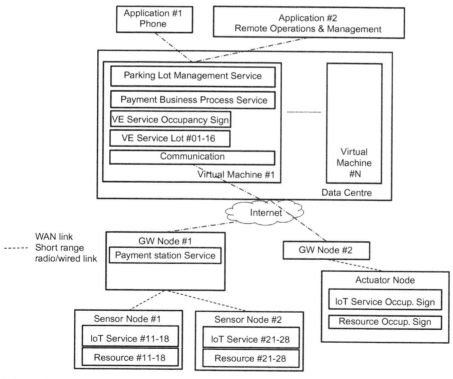

FIGURE 8.9

Parking Lot Deployment & Operational View, Resources, Services, Virtual Entities, Users.

Services appear in the figure because an IoT system is typically part of a larger system. Starting from the Sensor Devices, as seen earlier, Sensor Node #1 hosts Resource #11—#18, representing the sensors for the parking spots #01—#08, while earlier Sensor Node #2 hosts Resource #21—#28, representing the sensors for the parking spots #09—#16. We assume that the sensor nodes are powerful enough to host the IoT Services #11—#18 and #21—#28 representing the respective resources. The two sensor nodes are connected to the gateway device that also hosts the payment service with the accompanying sensors and actuators, as seen earlier. The other gateway device hosts the occupancy sign actuator resource and corresponding service. The management system for the specific parking lot, as well as others, is deployed on a virtual machine on a data center. The virtual machine hosts communication capabilities, Virtual Entity services for the parking spots #01—#16, the Virtual Entity services for the occupancy sign, a payment business process that involves the payment station and input from the occupancy sensor services, and the parking lot management service that provides exposure and access control to the parking lot occupancy data for the parking operation center and the consumer phone applications. As a reminder, the Virtual Entity service of the parking lot uses the IoT Services hosted on two sensor nodes and performs the mapping between the sensor node identifiers (#11—#18 and #21—#28) to parking spot identifiers (spot #01—#16). The services offered on these parking spots are to read the current state of the parking spot to see whether it is "free" or "occupied." The Virtual Entity corresponding to the occupancy sign contains one writable attribute: the number of free parking spots. A User writing this Virtual Entity attribute results in an actuator command to the real actuator resource to change its display to the new value.

Of course, for the operation of the parking lot, a number of other IoT-related services would be useful, such as historical occupancy data on which machine learning algorithms could support the operator of the parking lot in making decisions about planning and charging.

Starting from the IoT Domain Model, we attempt to perform a high-level mapping between the different classes/entities of the model and their realization.

The physical sensors, actuators, tags, processors, and memory, which are parts of a Device, are deployed close to the Physical Entities of Interest, the ones whose properties are monitored or controlled.

Figure 8.10 shows an example of mapping an IoT Domain Model and Functional View to Devices with different capabilities (different alternatives) connecting to a cloud infrastructure. Alternative 1 shows devices that can

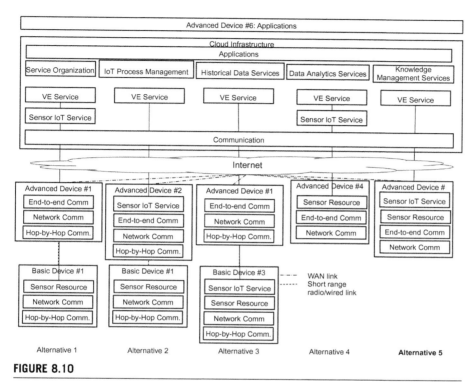

Advanced Device #6: Applications

Cloud Infrastructure
Applications

| Service Organization | IoT Process Management | Historical Data Services | Data Analytics Services | Knowledge Management Services |

VE Service | VE Service | VE Service | VE Service | VE Service

Sensor IoT Service | | | Sensor IoT Service |

Communication

Internet

Alternative 1 | Alternative 2 | Alternative 3 | Alternative 4 | **Alternative 5**

Advanced Device #1
- End-to-end Comm
- Network Comm
- Hop-by-Hop Comm.

Advanced Device #2
- Sensor IoT Service
- End-to-end Comm
- Network Comm
- Hop-by-Hop Comm.

Advanced Device #1
- End-to-end Comm
- Network Comm
- Hop-by-Hop Comm.

Advanced Device #4
- Sensor Resource
- End-to-end Comm
- Network Comm

Advanced Device #
- Sensor IoT Service
- Sensor Resource
- End-to-end Comm
- Network Comm

Basic Device #1
- Sensor Resource
- Network Comm
- Hop-by-Hop Comm.

Basic Device #1
- Sensor Resource
- Network Comm
- Hop-by-Hop Comm.

Basic Device #3
- Sensor IoT Service
- Sensor Resource
- Network Comm
- Hop-by-Hop Comm.

WAN link
Short range radio/wired link

FIGURE 8.10

Mapping IoT Domain Model concepts to Deployment View.

host only a simple Sensor Device and a short-range wired or wireless connectivity technology (Basic Device #1). Such kind of device needs an Advanced Device of type #1 that allows the basic device to perform protocol adaptation (at least from the short-range wired or wireless connectivity technology to a WAN technology) so that the Sensor IoT service in the cloud and the Sensor Resource on the Basic Device #1 can exchange information. The Virtual Entity representing the Physical Entity where the Basic Device #1 is deployed is also hosted in the cloud. In alternative 2, Advanced Devices (type #2) can host the Sensor IoT Service communicating to the Sensor Resource on a Basic Device #1. The cloud infrastructure in this case only hosts the Virtual Entity Service corresponding to the Sensor IoT Service. The difference between alternative 1 and 2 is that the Sensor IoT Service hosted on an Advanced Device #2 should be capable of responding to requests from Users (cloud services, Applications) with the appropriate secure mediation of course. In alternative 3, the Basic Device #3 is capable of providing the Sensor Resource and the Sensor IoT Service but still needs an Advanced

Device #1 to transport IoT service requests/responses/subscriptions/ notifications/publications to the Users in the cloud. According to experience, this kind of deployment scenario imposes a high burden on a Basic Device, which potentially makes the Basic Device the weakest link in the information flow. If malicious Users launch a Denial of Service (DoS) attack on the node, the probability of the node going down is very high. Alternatives 4 and 5 show Advanced Devices offering a WAN interface. In alternative 4, only the Sensor Resource is hosted on the Device, while in alternative 5, even the IoT Service is hosted on the Device. The Virtual Entity Service is hosted in the cloud.

The cloud infrastructure contains, apart from Virtual Entity services, Service Organization components (Composition, Orchestration, Choreography), IoT Process Management components, Historical Data Services (collection, processing), Data Analytics and Knowledge Management, to name a few. The list is by no means exhaustive.

Last but not least, applications can be on different types of devices or in the cloud. Advanced Devices of type #6 can host applications that use either local communication capabilities to exchange information to Basic or Advanced Devices attached to Physical Entities, or exchange information with the cloud infrastructure.

8.5 Other relevant architectural views

Apart from these functional views, there are a few more that are very important for a system that interfaces the physical world. The two most important are the Physical Entity View and the Context View. They are not covered here in detail because they are directly dependent on the actual IoT system properties, which vary from use case to use case.

The Physical Entity View describes the Physical Entities from the IoT Domain Model in terms of physical properties (e.g. dimensions for spaces/ objects). The description of the Physical Entities may also include the relationship between Physical Entities (e.g. an entity is included in the other and may be stationary in a specific location or mobile). The large number of possibilities for Physical Entities cannot be captured in a Reference Architecture, nevertheless, an architect needs to outline all the details of the Physical Entities from the beginning in order to assess if any Physical Property affects the rest of the architectural views and models.

According to Rozanski & Woods (2011), the Context of a system, "describes the relationships, dependencies, and interactions between the system and its environment (people, systems, external entities with which it interacts)." Therefore, the Context View should capture external entities interacting with the system, impact of the system on its environment, external entity properties/identities, system scope and responsibilities, etc. Since the possibilities for external entities and their interactions with an IoT system depend on the assumptions on the actual system, this view is also constructed in the beginning of the design process because it sets the boundary conditions for the problem at hand. In the parking lot example above, we described briefly parts of the Physical Entity and Context View without explicit individual presentations of these views. For example, the dimensions of the parking spots are a Physical Entity property, and the fact that there are sixteen parking spots physically placed in a possibly gated parking lot, the fact that there is an occupancy display near the parking lot on the roadside, and other details outline both Physical Entity properties as well as relationships between the system and its environment.

Real-World Design Constraints

9

CHAPTER OUTLINE

9.1 Introduction

This chapter outlines the technical design constraints to illustrate the questions that need to be taken into account when developing and implementing M2M and IoT solutions in the real world. This provides some background and thoughts for the use cases outlined in Part III of this book.

9.2 Technical design constraints — hardware is popular again

The IoT will see additional circuitry built into a number of existing products and machines — from washing machines to meters. Giving these things an identity, and the ability to represent themselves online and communicate

From Machine-to-Machine to the Internet of Things: Introduction to a New Age of Intelligence.
DOI: http://dx.doi.org/10.1016/B978-0-12-407684-6.00009-7

with applications and other things, represents a significant, widely recognized opportunity.

For manufacturers of products that typically contain electronic components, this process will be relatively straightforward. Selection of appropriate communications technologies that can be integrated with legacy designs (e.g. motherboards) will be relatively painless. The operational environments and the criticality of the information transmitted to and from these products, however, will present some unconventional challenges and design considerations. These are discussed later in the context of new and potential applications.

The IoT will, on the other hand, allow for the development of novel applications in all imaginable scenarios. Emerging applications of M2M and wireless sensor and actuator networks have seen deployment of sensing capabilities in the wild that allow stakeholders to optimize their businesses, glean new insight into relevant physical and environmental processes, and understand and control situations that would have previously been inaccessible.

The technical design of any M2M or IoT solution requires a fundamental understanding of the specificity of the intended application and business proposition, in addition to heterogeneity of existing solutions. Developing an end-to-end instance of an M2M or IoT solution requires the careful selection, and in most cases, development of a number of complementary technologies. This can be both a difficult conceptual problem and integration challenge, and requires the involvement of the key stakeholder(s) on a number of conceptual and technological levels. Typically, it can be considered to be a combinatorial optimization problem — where the optimal solution is the one that satisfies all functional and nonfunctional requirements, whilst simultaneously delivering a satisfactory cost-benefit ratio. This is particularly relevant for organizations wishing to compete with existing offerings, or for start-up ventures in novel application areas. Typically, capital costs in terms of "commissioning" and operational costs in "maintenance" must be considered. These may be balanced by resultant optimizations.

Typical M2M or IoT applications conform to the general functional architecture presented in Chapters 6, 7, and 8. Assuming that the system designer has selected the appropriate communications technologies to bridge the device and application domains (likely standard Internet Protocol (IP)-based methods as described in Chapter 5), he or she must consider the application at several levels: the device (or M2M Area Network; i.e. hardware), representation (i.e. data and visualization thereof), and interaction (i.e. local or remote control).

9.2.1 Devices and networks

Introduced in Chapter 5, devices that form networks in the M2M Area Network domain must be selected, or designed, with certain functionality in mind. At a minimum, they must have an energy source (e.g. batteries, increasingly EH), computational capability (e.g. an MCU), appropriate communications interface (e.g. a Radio Frequency Integrated Circuit (RFIC) and front end RF circuitry), memory (program and data), and sensing (and/or actuation) capability.

These must be integrated in such a way that the functional requirements of the desired application can be satisfied, in addition to a number of non-functional requirements that will exist in all cases.

9.2.1.1 Functional requirements

Specific sensing and actuating capabilities are basic functional requirements. In every case – with the exception of devices that might be deployed as a routing device in the case of range issues between sensing and/or actuating devices – the device must be capable of sensing or perceiving something interesting from the environment. This is the basis of the application. Sensors, broadly speaking, are difficult to categorize effectively. Selecting a sensor that is capable of detecting a particular phenomenon of interest is essential. The sensor may directly measure the phenomenon of interest (e.g. temperature), or may be used to derive data or information about the phenomenon of interest, based on additional knowledge (e.g. a level of comfort). Sensors may sense a phenomenon that is local (i.e. a meter detecting total electricity consumption of a space) or distributed (e.g. the weather).

In many cases, sensing may be prohibitively expensive or unjustifiable at scale, and thus motivates the derivation of models that can reason over the sensor readings that are available. Air and water quality monitoring systems are typical of this type of problem.

Given a particular phenomenon of interest, there are often numerous sensors capable of detecting the same phenomenon (e.g. types of temperature sensors), but have widely varying characteristics. These characteristics relate to the accuracy of the sensor, its susceptibility to changing environmental conditions, its power requirements, its signal conditioning requirements, and so forth.

In some cases, for example, a complementary (e.g. temperature) sensor is required in addition to the primary sensor such that variations in readings of the primary sensor that are caused by variation in temperature can be understood in context.

Sensing principle and data requirements are also of essence when considering the real-world application. Consider a continuously sampling sensor, such as an accelerometer, versus a displacement transducer. Displacement can be sampled intermittently, whereas if an accelerometer is duty-cycled, it is likely that data points of interest (i.e. real-world events) may be missed. Furthermore, the data requirements of the stakeholder must be taken into account. If all data points are required to be transmitted (which is the case in many scenarios, irrespective of the ability to reason locally within an M2M Area Network or WSN), this implies higher network throughput, data loss, energy use, etc. These requirements tend to change on a case-by-case basis.

9.2.1.2 Sensing and communications field
The sensing field is of importance when considering both the phenomenon to be sensed (i.e. Is it local or distributed?) and the distance between sensing points. The physical environment has an implication on the communications technologies selected and the reliability of the system in operation thereafter. Devices must be placed in close enough proximity to communicate. Where the distance is too great, routing devices may be necessary. Devices may become intermittently disconnected due to the time varying, stochastic nature of the wireless medium. Certain environments may be fundamentally more suited to wireless propagation than others. For example, studies have shown that tunnels are excellent environments for wireless propagation, whereas, where RF shielding can occur (e.g. in a heavy construction environment), communication range of devices can be significantly reduced.

9.2.1.3 Programming and embedded intelligence
Devices in the IoT are fundamentally heterogeneous. There are, and will continue to be, various computational architectures, including MCUs (8-, 16-, 32-bit, ARM, 8051, RISC, Intel, etc.), signal conditioning (e.g. ADC), and memory (ROM, (S/F/D)RAM, etc.), in addition to communications media, peripheral components (sensors, actuators, buttons, screens, LEDs), etc. In some applications, where it would previously have been typical to have homogeneous devices, a variety of sensors and actuators can actually exist, working collaboratively, but constituting a heterogeneous network in reality.

In every case, an application programmer must consider the hardware selected or designed, and its capabilities. Typically, applications may be thought of cyclically and logically. Application-level logic dictates the sampling rate of the sensor, the local processing performed on sensor readings, the transmission schedule (or reporting rate), and the management of the communications protocol stack, among other things. Careful implementation of the (embedded) software is required to ensure that the device operates as desired. This continues to be non-trivial and highly specialized. For heterogeneous devices, the embedded software will vary by device.

The ability to reconfigure and reprogram devices is still an unresolved issue for the research community in sensor networks, M2M, and the IoT. It relates both to the physical composition of devices, logical construction of the embedded software, and addressability of individual devices and security, to name a few. Operating systems are typically used to make programming simpler and modular for embedded systems designers, but each comes with conceptual and implementation differences that impact the ability to handle certain desirable features.

9.2.1.4 Power

Power is essential for any embedded or IoT device. Depending on the application, power may be provided by the mains, batteries, or conversion from energy scavengers (often implemented as hybrid power sources). The power source has a significant implication on the design of the entire system. If a finite power supply is used, such as a battery, then the hardware selected, in addition to the application level logic and communications technology, collectively have a major impact on the longevity of the application. This results in short-lived applications or increased maintenance costs. In most cases, it should be possible to analytically model the power requirements of the application prior to deployment. This allows the designer to estimate the cost of maintenance over time.

9.2.1.5 Gateway

The Gateway, described in Chapter 5, is typically more straightforward to design if it usually acts as a proxy; however, there are very few effective M2M or IoT Gateway devices available on the market today. Depending on the application, power considerations must be taken into account. It is also thought that the Gateway device can be exploited for performing some level of analytics on data transitioning to and from capillary networks.

9.2.1.6 Nonfunctional requirements

There are a number of nonfunctional requirements that need to be satisfied for every application. These are technical and non-technical:

- Regulations
  - For applications that require placing nodes in public places, planning permission often becomes an issue.
  - Radio Frequency (RF) regulations limit the power with which transmitters can broadcast. This varies by region and frequency band.

- Ease of use, installation, maintenance, accessibility
  - Simplification of installation and configuration of IoT applications is as yet unresolved beyond well-known, off-the-shelf systems. It is difficult to conceive a general solution to this problem. This relates to positioning, placement, site surveying, programming, and physical accessibility of devices for maintenance purposes.
- Physical constraints (from several perspectives)
  - Can the additional electronics be easily integrated into the existing system?
  - Are there physical size limitations on the device as a result of the deployment scenario?
  - What kind of packaging is most suitable (e.g. IP-rated enclosures for outdoor deployment)?
  - What kind and size of antenna can I use?
  - What kind of power supply can I use given size restrictions (relates to harvesting, batteries, and alternative storage, e.g. supercapacitors)?

9.2.1.7 Financial cost

Financial cost considerations are as follows:

- **Component Selection:** Typically, the use of these devices in the M2M Area Network domain is seen to reduce the overall cost burden by using non-leased communications infrastructure. However, there are research and development costs likely to be incurred for each individual application in the IoT that requires device development or integration. Developing devices in small quantities is expensive.
- **Integrated Device Design:** Once the energy, sensors, actuators, computation, memory, power, connectivity, physical, and other functional and nonfunctional requirements are considered, it is likely that an integrated device must be produced. This is essentially going to be an exercise in Printed Circuit Board (PCB) design, but will in many cases require some consideration to be paid to the RF front-end design. This means that the PCB design will require specific attention to be paid to the reference designs of the RFIC manufacturer during development, or potentially the integration of an additional Integrated Circuit (IC) that deals with the balun and matching network required.

9.3 Data representation and visualization

Each IoT application has an optimal visual representation of the data and the system. Data that is generated from heterogeneous systems has heterogeneous visualization requirements. There are currently no satisfactory standard data representation and storage methods that satisfy all of the potential IoT applications.

Data-derivative products will have further *ad hoc* visualization requirements. A derivative in these terms exists once a function has been performed on an initial data set − which may or may not be raw data. These can be further integrated at various levels of abstraction, depending on the logic of the integrator. New information sources, such as those derived from integrated data streams from various logically correlated IoT applications, will present interesting representation and visualization challenges.

9.4 Interaction and remote control

To exploit remote interaction and control over IoT applications, connectivity that spans the traditional Internet (i.e. from anywhere) on the side of the application manager, or other authorized entity, to the end-point (i.e. an embedded device), continues to be a challenging problem. Aside from authentication and availability challenges, for most constrained devices, heterogeneous software architectures, such as event-based operating systems running on devices with significantly varying concurrency models, continue to pose significant challenges from a remote management perspective.

Elements of Device Management, specifically reprogramming and reconfiguration of deeply embedded devices, will be required, particularly for devices deployed in inaccessible locations. This requires, among others, reliability, availability, security, energy efficiency, and latency performance, to be satisfactory whilst communicating across complex distributed systems.

Another significantly under-researched topic is the definition and delivery of end-to-end quality of service (QoS) metrics and mechanisms in IoT-type applications. These will be necessary if Service Agreements (SA) or Service Level Agreements (SLA) are to be defined in the case of service provisions for IoT applications − which may or may not be desirable to the application owner. This will be situation-specific. End-to-end latency, security, reliability, availability, times between failure and repair, responsibility, etc., are all likely to feature in such agreements.

IoT Use Cases

The following chapters provide an overview of some of the main use cases for M2M and IoT.

Industrial automation systems constitute the backbone of modern society as they use control systems and Information Technologies (IT) to guarantee a multitude of tasks, such as the production of goods and delivery of services, and generally monitor and control much of the processes upon which the global economy is based.

Industrial automation systems constitute complex ecosystems that amalgamate physical machines with modern information and communication technologies and human interaction (where applicable) in order to operate efficiently in complex processes. Typical components involved in automation systems include Distributed Control Systems (DCS), Supervisory Control and Data Acquisition Systems (SCADA), Programmable Logic Controllers (PLC), Sensors, Robotics, Management Systems, and Enterprise Systems, just to name a few. However, industrial automation is directly or indirectly

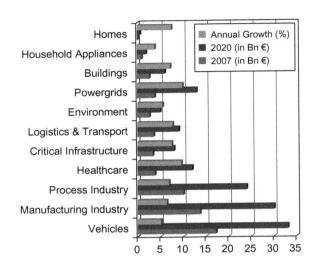

(European Commission 2008)

FIGURE III.1

Monitoring and Control Market 2007−2020.

involved in several domains with significant footprints, as shown in Figure III.1.

M2M in monitoring and control is seen as an integral part of Industrial Automation Systems, with a significant impact on the global economy. As a European Commission (2008) study attests (Figure III.1), the world monitoring and control market is expected to grow from €188 billion in 2007, by €300 billion, reaching €500 billion in 2020. Between 2007 and 2020 the European monitoring and control domain is expected to grow at a 5.7% rate annually. With a share of €61.5 billion, Europe represents 32% of this market. Services, with more than 50% of the market value, have the biggest share. Together, three application markets − Vehicles, Manufacturing, and Process industries − represent 60% of the total monitoring and control market. Healthcare, critical infrastructures, and logistic and transport follow closely, while at the moment, the Home is still considered a small niche market.

M2M is expected to play a key role and provide the ability to realize new approaches that are cost and energy efficient, while in parallel are flexible and open to future innovations. Integration of M2M for monitoring and control is not expected to be trivial, especially in an environment that is used to isolated tight control of the infrastructure and a pace of evolution much slower. Industrial systems pose significantly different hard requirements that need to be tackled, especially when related to critical infrastructures. We will investigate here how the transition of existing factories to future ones with the use of service-oriented technologies and the emerging cloud can be realized.

Future factories are expected to be complex System of Systems (SoS) that will empower a new generation of applications and services that are today hardly realizable or too costly. New sophisticated enterprise-wide monitoring and control approaches will be possible due to the prevalence of Cyber-Physical Systems (CPS), which have made M2M interactions a key competitive advantage and market differentiator. This will be possible due to several disruptive advances, as well as the cross-domain fertilization of concepts and the amalgamation of IT-driven approaches in traditional industrial automation systems. The Factory of the Future (FoF) will rely on a large ecosystem of systems where collaboration at large-scale will take place.

Similar systems based on largely identical technology and ideology will be the basis of the technological core of Smart Cities and the Smart Grid (and other Smart "Utilities"), coupled with increasing engagement with participatory sensing (PS) systems and more traditional enterprise computing infrastructure.

Let's see how some of these use cases are implemented in the following chapters.

Asset Management

10

CHAPTER OUTLINE

10.1 Introduction

The emergence of the IoT with its billions of envisioned devices poses a clear challenge to the management of these. Existing asset management practices consider the operations applicable to various physical assets, and in their majority refer to monitoring of their operations and to some degree to the adjustment (control) of their behavior. However, such operations have been up to now strongly coupled with what the devices and underlying systems are capable of, bound with (mostly) proprietary protocols in order to cover the largest possible spectrum of functionalities and guarantee results, and are mostly static.

All of these current practices, however, are undergoing transformation. The reasons include the explosion of the heterogeneity of devices that are now deployed in modern infrastructures, which can deliver high-quality data that have clear business relevance at a fraction of the cost compared to some years ago. The usage of open standards is one promising way to go to achieve large-scale manageability, however, it is not going to be enough. The reasons lie in the increasing complexity not only of the devices themselves, but also on the constellations they take part in, and how their functionalities are used in modern applications. On top of this, applications are not monolithic from one provider, but on the contrary depend on multiple layers that are developed from other stakeholders. Hence, managing different

From Machine-to-Machine to the Internet of Things: Introduction to a New Age of Intelligence.
DOI: http://dx.doi.org/10.1016/B978-0-12-407684-6.00010-3

devices requires significant overhead in order to be able to effectively integrate, monitor, and control/reconfigure them.

A typical manageability nightmare scenario includes the configuration of assets in modern enterprises. For example, today employees use several computing devices (e.g. laptops, desktops, smart phones, tablets, access cards, security tokens, etc.); all of these need to be accounted for in backend systems, to be integrated with enterprise-wide monitoring solutions, and to comply with the organization's policies and requirements. How to achieve that is challenging; for example, how does one protect a company's assets and the data they contain from unauthorized access or usage? The latter also comes with, for example, a trade-off between security and functionality in order to enhance the employees' performance and benefits from using the devices, while still adhering to the overall enterprise constraints.

10.2 Expected benefits

The IoT era is dominated with interactions among the devices, access to their data, and dynamic configuration/management of them. The prevalence of modern Information Technology (IT) concepts and adoption of Internet-based technologies is slowly penetrating traditional domains such as those of energy, manufacturing, etc. Hence, management in the IoT era, although challenging, may yield significant benefits and enable the mastering of the vast device-based infrastructure.

Several benefits are expected with M2M in asset management. These may include:

- Reduced costs, e.g. because of remote telemetry without the need of field personnel to be engaged.
- Increased quality, e.g. because of the fine-grained monitoring data that could be done even in near real-time.
- Increased resilience, e.g. because of analysis of device's status, can lead to predictive maintenance, which minimizes apart from costs, and also downtime and unexpected failures.
- Increased performance and security: remote updates may enhance the operational capabilities of the assets.
- Increased security: updates of the asset's software can help correct its behavior and security holes.
- Asset location tracking, e.g. easier recovery and theft prevention of assets.

- Operation optimization, e.g. fleet management optimizing for journeys on the fly.
- New services, e.g. energy awareness via smart metering, location-based services, e-ticketing, etc.

On the downside, if all of the benefits rely on the correct tackling of complex issues, then only partially fulfilling them might quickly turn an advantage into a disadvantage. As an example, remote access to the devices lowers integration costs, but also potentially opens the door for third parties to tamper with them and their functionalities, which may lead to security risks, etc.

Asset management is a great challenge, but with pivotal benefits in the IoT era. Its applications are expected to empower the next generation of innovation in multiple domains, such as residential homes, healthcare, buildings, cities, transportation networks, energy grids, manufacturing, supply chains, homeland security, workplace safety, environmental monitoring, etc., just to name a few.

10.3 e-Maintenance in the M2M Era

The recent focus is on e-maintenance, i.e. "maintenance support which includes the resources, services, and management necessary to enable proactive decision process execution" (Muller et al. 2008), especially due to the expected benefits of M2M that can be harnessed. However, in order to effectively implement e-maintenance applications for asset lifecycle management, several requirements need to be met, and this is a challenging task. Apart from interoperability, a major roadblock is the absence of IT platforms (Cannata et al. 2009) that master the new challenges and can fully support e-maintenance practices.

Key strategies (Muller et al. 2008, Cannata et al. 2009) of e-maintenance are the following:

- **Remote Maintenance**: The capability empowered by Information and Communication Technologies (ICT) to provide maintenance practices from anywhere without being physically present (e.g. third party entities outside the enterprise borders). This approach enables far more effective reaction to maintenance, and dramatically affects business models having maintenance services as part of their functionalities.
- **Predictive Maintenance**: Here the adoption of models and methodologies is implied that analyzes the operational performance

of assets and attempts to predict malfunctions and failures, or determine on-demand when maintenance checks are needed (and not as usually done at fixed intervals). Here, apart from the immediate benefit of asset reliability, one can also witness optimization of maintenance schedules carried out by field personnel, enhanced planning for replacement of assets, increased customer satisfaction, etc.

- **Real-Time Maintenance**: Current failures on the shop floor, although evident, might require significant time until they are assessed, repaired, and re-planned with respect to the operations the enterprise systems had scheduled. With real-time notifications up to enterprise systems, immediate assessment on the operational factory or enterprise-wide processes can be achieved and addressed.
- **Collaborative Maintenance**: This capability enables traditional maintenance concepts to integrate collaboration among different areas of the enterprise that may lead to leaner processes, simpler landscapes, and more effective management.

To achieve these objectives, we have to harness the new capabilities that M2M brings. As M2M approaches increasingly integrate Internet technologies that have been proven to be open enough (compared to many practices in individual industry domains) and scalable, it is expected that integration overall will be easier considering the large heterogeneity of assets. Since the assets are no longer passively monitored entities within an infrastructure, they can take advantage of the event-driven methods present in M2M and on-demand report their health status, failures, and/or other information. This migration from a traditional information-pull to an event-driven approach is expected to minimize unnecessary traffic and empower the targeted information dissemination within the infrastructure.

Another key aspect of M2M that e-maintenance can benefit from is the dynamic discovery of assets and their "surrounding context" information such as location or real-world sensing information of the process they monitor or control. Integration of protocols that allow a simple "plug & play" approach mean that assets can be connected, immediately recognized by the network services, get automatically configured, and operate within the organization's policies.

However, apart from these example asset-related enhancements that M2M brings, collaborative e-maintenance is a promising approach that integrates multi-enterprise experts that can together effectively address any asset issues. This is of critical importance and paves the way for new business models deploying subject matter experts when and where needed remotely, which leads to more effective usage of the expertise and better solutions to the maintenance problems (Cannata et al. 2009).

Cross-company communication is already a reality, but constrained at enterprise-level operations without in-depth information that could trigger sophisticated re-planning scenarios. However, with M2M real-time connection to the devices, malfunctions can be quickly analyzed and resolved with the help of multiple experts such as those with knowledge of the specific process, those responsible for the hardware/software (HW/SW) of the device, and potentially the field personnel on-site (as depicted in Figure 10.1). All these can now collaborate using multiple Internet-based technologies with seamless data, voice, and video integration over a future e-maintenance platform.

Communication done directly (e.g. via common trusted third-party service providers) may simply couple the two companies for the specific business case, and remove the overhead of costly home-grown solutions by propagating all the necessary information to the affected stakeholders. Synergies can be identified, and information that was too costly to be obtained in a timely

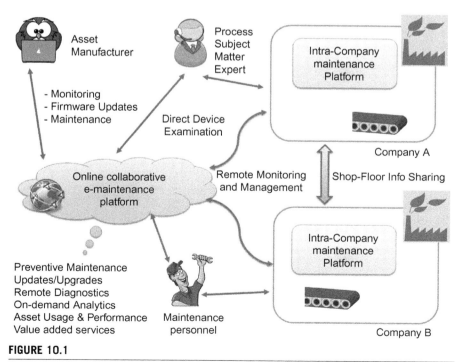

FIGURE 10.1

Outsourced remote continuous cross-company e-maintenance.

manner can now flow into cross-company applications and services. This approach is very well-suited for dynamic and short-lived interactions that can be set up, exploited, and removed as easy as a simple composite service.

Cross-company collaboration allows us to realize new functionality and innovate on services offered (Cannata et al. 2009). Especially in the case of outsourcing of maintenance, specialized partners can now bring in their expertise and monitor remotely the devices on the shop floor and maintain them. Assets on which the company operates may in the future not be owned by the company as such, but instead be provided to them over specific service level agreements (SLAs), e.g. a production line with uptime of 99%. How this is achieved and maintained is the responsibility of the service provider. As a result, companies can now focus more on their core business, while SLAs can regulate shop-floor performance that better matches the business process goals, but not how this is achieved, which is the responsibility of the e-maintenance partner. This can facilitate the development of new business models based on remote maintenance service delivery through e-maintenance platforms.

10.4 Hazardous goods management in the M2M Era

An example of asset management that goes beyond the traditional monitoring approaches is that realized within the research project Collaborative Business Items (CoBIs 2013). There, an M2M platform was used to monitor hazardous goods (i.e. chemicals) and guarantee their safe storage. Storage in close proximity of incompatible goods i.e. goods that could be flammable if brought together, or exceeding the compliance guidelines, would raise alarms locally and at the enterprise system that would need to be resolved (Karnouskos & Haller 2007).

Traditionally, the monitoring was done with passive radio frequency identification (RFID) tags, however, CoBIs deployed Wireless Sensor Networks (WSNs) that could execute business logic locally and communicate among themselves as well as with the enterprise systems (Decker et al. 2006). WSNs are seen as one of the most promising technologies that will bridge the physical and virtual worlds, enabling them to measure, assess, and actuate real-world environments.

A typical demonstration of how M2M helps with the hazardous goods management scenario is shown in Figure 10.2. All drums are equipped with WSNs that have internally stored information about the chemical within the drum, an "incompatible goods" list indicating which chemicals

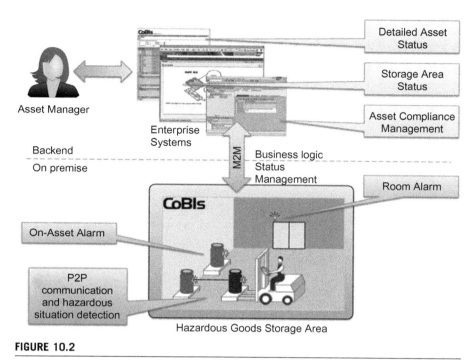

FIGURE 10.2

M2M-enhanced Hazardous Goods Management in CoBIs.

are dangerous to be in near proximity with it as corrosion might lead to explosions, etc., as well as the max limit according to current regulations of this chemical that can be stored within a limited space. The WSNs are able to beacon other WSNs nearby and exchange information about the chemicals within the drum. A nearby WSN bridge interacts with the WSNs of the drums and has a connection to an enterprise system, and hence mediates all information from the field (the drum storage facility). The WSN bridge is also capable of not only providing the backend with WSN information (bottom-up communication), but also providing other WSNs with information from the backend (top-down communication). The latter is handy, as information provided initially to WSNs may change; for example, the chemical incompatibility list may be extended or the max storage limits may be increased or reduced. The WSNs and the bridge also feature visible and audible alarms.

In a typical safety-critical situation, a worker might transfer by mistake a drum and position it on the wrong side of the storage room (i.e. near another drum whose potential combination may cause explosions). As the

WSNs scan their neighbors and exchange information, the dangerous situation is immediately detected by the WSNs of both the drum that is being transferred as well as the other drums already in that location (due to the incompatibility list they host locally). An alarm goes off on the drums and the storage room, and the worker resolves the situation on-site by moving it away to the correct location. The same alarm is also visible on the enterprise systems for remote asset monitoring managers who could also take additional actions. Without the M2M interaction, such a situation would probably go undetected and impact the safety of the storage room workers. A similar hazardous situation may arise if more than the allowed drums with chemicals are stored in a single location (in violation of legislation). As an additional drum is placed, similarly the excess of the storage limit is identified and an alarm goes off. The backend of course has additional info on how to resolve the situation. Changes in the limits of max storage can be pushed wirelessly to the WSNs to keep the whole system in sync.

The real-time interaction at "the point of action" (i.e. on-premise by M2M) has profound benefits. The infrastructure empowered by IoT can react to events locally, and the field personnel can take corrective actions. The latter can also be realized without any connection to the enterprise systems because the logic of detection of hazardous situations is locally hosted on WSNs (and updatable by the enterprise system). The connection with the enterprise systems offers additional benefits linking asset management processes with the real-time status of the field, effectively enabling remote monitoring and actuation scenarios related to asset management.

10.5 Conclusions

Asset management is an area that can hugely benefit from M2M. New innovative solutions can be realized that take advantage of the networked embedded devices on-premise, the information they provide, and the collaboration with enterprise systems. However, for such solutions to be adopted, several challenges, such as complexity management, interoperability, security, and quality of service (QoS)-guaranteed communication have to be tackled, especially when scenarios involving critical infrastructures are involved. Nevertheless, there is a huge potential for many diverse domains where new business models and opportunities will arise with asset management.

Industrial Automation

11

CHAPTER OUTLINE

11.1 Service-oriented architecture-based device integration

The emerging approach in industrial environments is to create system intelligence by a large population of intelligent, small, networked, embedded devices at a high level of granularity, as opposed to the traditional approach of focusing intelligence on a few large and monolithic applications. This increased granularity of intelligence distributed among loosely coupled intelligent physical objects facilitates the adaptability and re-configurability of the system, allowing it to meet business demands not foreseen at the time of design, and providing real business benefits (Karnouskos et al. 2009a).

The Service-Oriented Architecture (SOA) paradigm can act as a unifying technology that spans several layers, from sensors and actuators used for monitoring and control at shop-floor level, up to enterprise systems and their processes as envisioned in Figure 11.1. This common "backbone" means that M2M is not limited to direct (e.g. proximity) device interaction, but includes a wide range of interactions in a cross-layer way with a variety of heterogeneous devices, as well as systems and their services. This yields multiple benefits for all stakeholders involved. Such visions have been proposed (Colombo & Karnouskos 2009) and realized, demonstrating the benefits and challenges involved (IBM 2008, Karnouskos et al. 2009a, 2009b). Internet Protocol (IP)-based, and more specifically web technologies and protocols (e.g. OPC-UA, DPWS, REST, Web Services (WS), etc.), constitute a promising approach (Cannata et al. 2010) towards the fundamental goal of enabling easy integration of device-level services

From Machine-to-Machine to the Internet of Things: Introduction to a New Age of Intelligence.
DOI: http://dx.doi.org/10.1016/B978-0-12-407684-6.00011-5

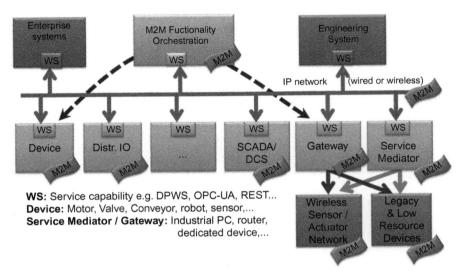

FIGURE 11.1

M2M SOA-based integration.

with enterprise systems overcoming the heterogeneity and specific implementation of hardware and software of the device. Surely industry specific requirements for security, resilience, and availability of near real-time event information needs to be effectively tackled. The latter are also seen as key enablers for the new generation of enterprise system applications such as business activity monitoring, overall equipment effectiveness optimization, maintenance optimization, etc.

The SOA-based vision is not expected to be realized overnight, but may take a considerable time depending on the lifecycle processes of the specific industry, and may be impacted by micro- and macro-economic aspects. Hence, it is important that migration capabilities are provided so that we can harvest some of the benefits today and provide a stepwise process towards achieving the vision. The concepts of gateway and service mediator (Karnouskos et al. 2009a), as depicted in Figure 11.2, can help towards this direction. Dynamic device discovery is a key functionality in the future M2M. As an example, Figure 11.3 depicts how Windows 7 can discover dynamically heterogeneous devices that are SOA-ready (i.e. equipped with web services; Devices Profile for Web Services, DPWS).

A Gateway is a device that controls a set of lower-level non-service-enabled devices, each of which is exposed by the Gateway as a service-enabled device. This approach allows gradually replacing limited-resource devices or legacy devices by natively WS-enabled devices without impacting the applications using these devices. This is possible since the same WS interface is offered this time by the WS-enabled device and not by the Gateway. This approach is used when each

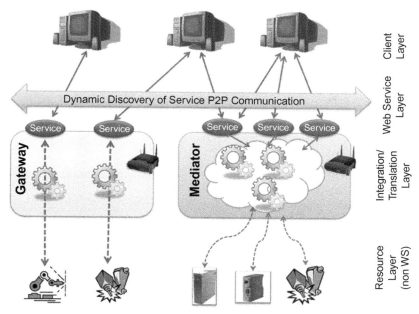

FIGURE 11.2

Non-service-enabled device integration: Gateway vs. Service Mediator.

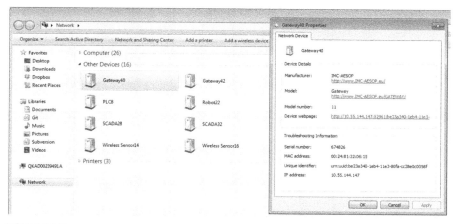

FIGURE 11.3

Dynamic device discovery via DPWS in Windows 7.

of the controlled devices needs to be known and addressed individually by higher-level services or applications.

Originally meant to aggregate various data sources (e.g. databases, log files, etc.), the Mediator components evolved and are now used to not only aggregate various services, but possibly also compute/process the data they acquire before exposing it as a service. Service Mediators aggregate, manage, and eventually represent services based on some semantics (e.g. using ontologies). In our case, the Service Mediator could be used to aggregate various non WS-enabled devices. This way, higher-level applications could communicate to Service Mediators offering WS instead of communicating to devices with proprietary interfaces. The benefits are clear, as we don't have the hassle of (proprietary) driver integration. Furthermore, now processing of data can be done at Service Mediator level, and more complex behavior can be created which was not possible before from the standalone devices.

As we can see in future IoT infrastructures dominated by billions of devices with different capabilities and needs, we have to consider how these integrate with each other and enable the realization of new innovative approaches. This assumes increased integration and collaboration among the various layers existing in industries (i.e. from the shop floor up to enterprise systems). Several concepts and efforts are directed towards abstracting from the device-specific aspects and defining device-agnostic, but functionality-focused, layers of integration.

11.2 SOCRADES: realizing the enterprise integrated Web of Things

Agility and flexibility are required from modern factories. This, in conjunction with the rapid advances in Information Technology (IT), both in hardware and software, as well as the increasing level of dependency on cross-factory functionalities, sets new challenging goals for future factories. The latter are expected to rely on a large ecosystem of systems where collaboration at large scale will take place. Mashing up services has proven to be a key advantage in the Internet application area; and if now the devices can either host web services natively or be represented as such in higher systems, then existing tools and approaches can be used to create mash-up apps that depend on these devices.

A visionary project that followed this line of thinking was the Industry-driven European Commission funded project SOCRADES (Colombo & Karnouskos 2009, Karnouskos et al. 2010a). Driven by the key need for cross-layer M2M collaboration (i.e. at shop-floor level among various heterogeneous devices as well as among systems and services up to the Enterprise (ERP) level), an architecture had been proposed, prototyped (Colombo et al. 2010), and assessed (Karnouskos et al. 2010a). SOCRADES proposed and realized SOA-based integration, as shown in Figure 11.2, including migration of existing infrastructure via gateways and service mediators, as shown in Figure 11.3.

To do so, it had to rely on an IoT architecture (depicted in Figure 11.3), whose primary goal was not the Peer-to-Peer (P2P) interaction among devices but their interaction with enterprise systems as well as the interactions among them, but with the assist of an infrastructure whose services could be utilized to enhance M2M operations (Karnouskos et al. 2010a).

The SOCRADES Integration Architecture (SIA), as analysed by Karnouskos et al. (2010a), enables enterprise-level applications to interact with and consume data from a wide range of networked devices using a high-level, abstract interface that features WS standards (Figure 11.4). One can clearly distinguish the various levels such as:

- **Application Interface**: This part enables the interaction with traditional enterprise systems and other applications. It acts as the glue for integrating the industrial devices, and their data and functionalities with enterprise repos and traditional information stores.
- **Service Management**: Functionalities offered by the devices are depicted as services here to ease the integration in traditional enterprise landscapes. Tools for their monitoring are provided.
- **Device Management**: Includes monitoring and inventory of devices, including service lifecycle management.

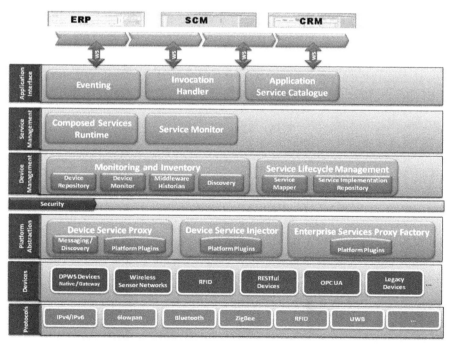

FIGURE 11.4

The SOCRADES Integration Architecture (SIA) enabling the coupling of (industrial) machines at shop-floor and enterprise systems.

- **Platform Abstraction**: This layer enables the abstraction of all devices independent of whether they natively support WS or not, to be wrapped and represented as services on the higher systems. In addition to service-enabling the communication with devices, this layer also provides a unified view on remotely installing or updating the software that runs on devices.
- **Devices & Protocols**: These layers include the actual devices that connect over multiple protocols to the infrastructure. The respective plugins of course need to be in place so that they can be seamlessly integrated to SIA.

To realize discovery and interaction in a P2P way, a local gateway/service-mediator is implemented. This prototype was named a Local Discovery Unit (LDU), and would enable the dynamic discovery of devices on premise and their coupling with the SIA.

SIA has been used in several scenarios as proof of concept for the integration among different devices, both locally and with enterprise systems. Examples include (Karnouskos et al. 2010a):

- Integration between a programmable logic controller, a robotic gripper, and SunSPOT wireless sensor nodes, while these are monitored by the SAP Manufacturing Integration and Intelligence software (SAP MII), which is also responsible for the execution of the business logic.
- Event-based interaction between Radio Frequency Identification (RFID) Reader (product ID via the RFID tag), robotic arm (used to demo transportation), a wireless sensor which monitored the usage of an emergency button, an IP-plugged emergency lamp, and a web application monitoring the actual production status and producing analytics.
- Production planning, execution, and monitoring via SAP MII of a test-rig controlled by Siemens Power Line Communication (PLC) and communicating over OPC.
- Passive energy monitoring via the usage of sensors (Ploggs) and gateways.

Although these are prototypes and not used productively, the proof of concept paves the way for further considerations towards using such approaches in real-world environments once the exact operational requirements are satisfied.

11.3 IMC-AESOP: from the Web of Things to the Cloud of Things

Visionary approaches have been further developed in the industry-driven project IMC-AESOP (2013). There the vision of SOCRADES has been pushed forward by considering the rapid advances in hardware and software, as well as IT concepts (Karnouskos & Colombo 2011). Therefore, we go beyond WS-enabled devices towards the cloud in order to harness its benefits, such as resource flexibility, scalability, etc. The result will be a highly dynamic flat information-driven infrastructure (Figure 11.5) that will empower the rapid development of better and more efficient next generation industrial applications, while in parallel satisfying the agility required by modern enterprises.

This vision (Karnouskos & Colombo 2011) is only realizable due to the distributed, autonomous, intelligent, pro-active, fault-tolerant, reusable (intelligent) systems, which expose their capabilities, functionalities, and structural characteristics as services located in a "service cloud." The infrastructure links many components (devices, systems, services, etc.) of a wide variety of scales, from individual groups of sensors and mechatronic components to, for example, whole control, monitoring and supervisory control systems, and performing SCADA, DCS, and MES functions (for example).

Although today factories are composed and structured by several systems viewed and interacting in a hierarchical fashion following mainly the specifications of standard enterprise architectures, there is an increasing trend to move towards information-driven interaction that goes beyond traditional hierarchical deployments and can coexist with them. With the empowerment offered by modern SOAs, the functionalities of each system (or even device) can be offered as one or more services of varying complexity, which may be hosted in the cloud and composed by other (potentially cross-layer) services, as depicted in Figure 11.5.

This transition marks a paradigm change in the interaction among the different systems, applications, and users. Although the traditional hierarchical view coexists, there is now a flat information-based architecture that depends on a big

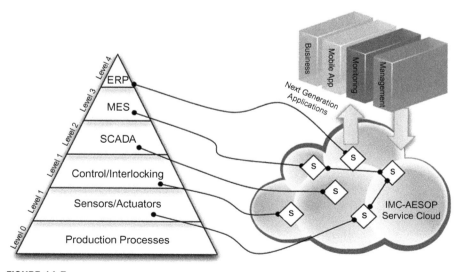

FIGURE 11.5

Future industrial system view of cloud-based composition of cyber-physical services.

variety of services exposed by the cyber-physical systems and their composition. Next generation industrial applications can now rapidly be composed by selecting and combining the new information and capabilities offered (as services in the cloud) to realize their goals. The envisioned transition to the future cloud-based industrial systems (Karnouskos & Colombo 2011, Karnouskos et al. 2012) is depicted in a Figure 11.6.

Several "user roles" will interact with the envisioned architecture (Figure 11.6), either directly or indirectly as part of their participation in a process plant. The roles define actions performed by staff and management, and simplifies grouping of tasks into categories such as business, operations, engineering, maintenance, engineering training, etc.

As depiced in Figure 11.6, it is possible to distinguish several service groups for which there have also been defined some initial services (Karnouskos et al. 2012). All of the services are considered essential, with varying degrees of importance for next generation, cloud-based, collaborative automation systems. The services are to provide key enabling functionalities to all stakeholders (i.e. other services, as well) as cyber-physical systems populating the infrastructure. As such, all these systems can be seen as entities that may have a physical part realized in on-premise hardware, as well as a virtual part realized in software potentially on-device and in-cloud. This emerging "Cloud of Things" (Karnouskos & Somlev 2013) has the potential to transform the way we design, deploy, and use applications and cyber-physical systems.

Typical functionalities include alarms, configuration and deployment, some control, data management, data processing, discovery, lifecycle management,

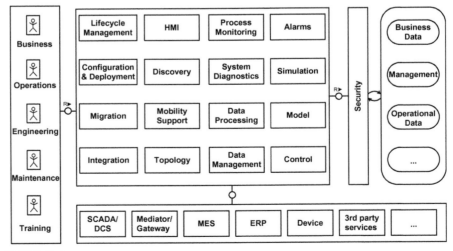

FIGURE 11.6

IMC-AESOP cloud-based architecture vision.

HMI, integration, simulation, mobility support, monitoring, security, etc. It is clear that this is a proposal that will need to be further refined in real-world scenarios, however, it clearly depicts a step towards a highly flexible M2M infrastructure for the automation domain that abstracts from devices and focuses on functionalities that can reside on-device, in-network, and harness the power of the cloud.

Outsourcing key functionalities to the cloud is challenging, and further research needs to be done towards the applicability of it for several scenarios. For instance, monitoring scenarios have been successfully realized demonstrating the benefits of the IT concept prevalence to traditional industrial system design and operation. However, several aspects with relation to real-time interactions, reliability, and resilience, as well as control (especially closed-loop control), are still at an early stage. Constructing such complex systems additionally bears challenges with respect to safety, maintainability, and security.

11.4 Conclusions

The prevalence of advanced embedded devices, in conjunction with increased processing and communication capabilities, is transforming industrial automation. Service-oriented approaches are now possible at device level, and bring with them benefits previously only available to the designers and managers of enterprise systems. We see a clear trend towards information-driven integration that empowers M2M interactions and integration with enterprise systems and the respective business processes. The latest visions depict a fully dynamically customizable infrastructure that harnesses the best breed of functionalities at device and at cloud level in order to empower the next generation of applications and services. Surely several challenges are open and need to be effectively tackled, however, on-going research will shed more light on the real-world applicability of such visions.

The Smart Grid

12

CHAPTER OUTLINE

12.1 Introduction

A revolution is currently underway in the electricity system, which has remained largely unchanged for more than 100 years. Rapid advances in Information Technology (IT) are increasingly integrated in several infrastructure layers covering all aspects of the electricity grid and its associated operations. In addition, intelligent networked devices are emerging whose M2M interactions create new capabilities in the monitoring and management of the electricity grid and the interaction between its stakeholders. IT-empowered innovations integrated with the electricity network and the stakeholders' interactions have paved the way (BDI 2010, Bryson & Gallagher 2012, European Commission 2012,) towards a "Smart Grid" that takes advantage of sophisticated bidirectional interactions.

The National Institute of Standards and Technology (NIST) Smart Grid Conceptual Model (Bryson & Gallagher 2012) defines a framework that outlines seven domains: Bulk Generation, Transmission, Distribution, Customers, Operations, and Markets and Service Providers. Complementary to that, the IEEE views the Smart Grid as "a large 'System of Systems' where each NIST Smart Grid domain is expanded into three Smart Grid foundational layers: (i) the Power and Energy Layer, (ii) the Communication Layer, and (iii) the IT/Computer Layer (Figure 12.1). Layers (i) and (iii) are

From Machine-to-Machine to the Internet of Things: Introduction to a New Age of Intelligence.
DOI: http://dx.doi.org/10.1016/B978-0-12-407684-6.00012-7

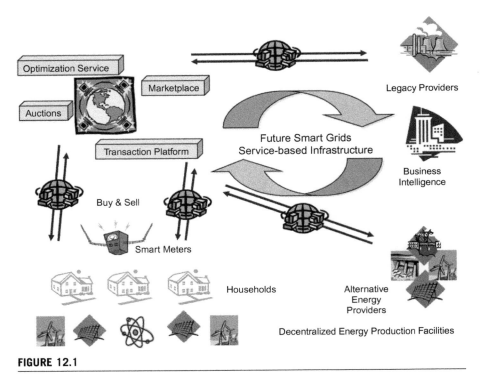

Optimization Service

Marketplace

Auctions

Legacy Providers

Transaction Platform

Future Smart Grids
Service-based Infrastructure

Business
Intelligence

Buy & Sell

Smart Meters

Households

Alternative
Energy
Providers

Decentralized Energy Production Facilities

FIGURE 12.1

The future ICT-empowered interaction-rich Smart Grid.

enabling infrastructure platforms of the Power and Energy Layer that makes the grid 'smarter'" (IEEE Smart Grid 2013).

The deregulation of the energy market and the subsequent new operational context will permanently change the structures associated with the production and distribution of energy (Karnouskos & Terzidis 2007). Deregulation or re-regulation of the energy market is designed to break up the value-added chain, with the production, transfer, and distribution of electric power forming separate segments. The objective is to establish a more open market designed to promote increased competition and flexibility for consumers. A much more decentralized and diversified energy production system will emerge as a result. New energy technologies for co-generated heat and power, and increased use of renewable energies such as bio-mass, solar energy, and wind power will introduce a considerable number of diversified systems into the power grid in addition to traditional large-scale plants. Consequently, the share of decentralized generated power — produced by industrial or private producers — will increase significantly. This will create

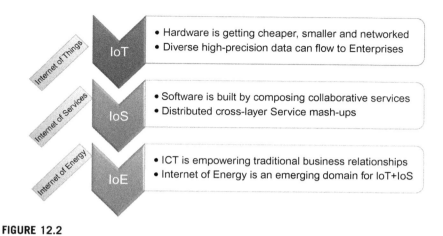

FIGURE 12.2

The Smart Grid technology based on a combination of IoT and IoS empowering old and new innovative energy-related interactions.

a new infrastructure where the future user may not be a simple passive consumer as today, but also a producer.

The integration of small, highly distributed energy production sources and their coupling with advanced, information-driven services will give rise to a new infrastructure (i.e. the Internet of Energy) (Karnouskos & Terzidis 2007), whose key technology building blocks are the IoT and Internet of Services (IoS). As shown in Figure 12.2, IoT and IoS make possible diverse M2M interactions locally and over the Internet, empowering traditional business relationships with fine-grained monitoring and control of the large energy infrastructure. The result is the Internet of Energy, which is the vision of the Smart Grid, spanning not only the technical grid infrastructure, but covering all energy-related aspects of the grid, its devices (including appliances, etc.), their interactions, as well as high-level applications and systems that depend on their data.

The Smart Grid is one of the key areas where the economic benefits of monitoring and control are likely to be realized (European Commission 2008). It is driven by "the degree of decentralization of the system components and their interrelation with electricity networks, the variability of renewable generation, the increased distance between electricity generation and consumption, the intelligence level of the involved systems created by smart products and associated smart services, the legal framework, the associated regulation of market-based product and service choices versus natural monopoly products and services and the business roles for actors involved in all aspects of networks and intelligent electric systems" (European Commission 2012). Currently, several projects (Giordano et al. 2013) with varying degrees of maturity are addressing several aspects of the Smart Grid value chain, investigating the benefits for the diverse stakeholders.

Due to the complexity of the multiple Smart Grid stakeholder interactions at several layers, it is mandatory to look at the Smart Grid from the network viewpoint, as an ecosystem where collaboration and information-driven interactions characterize it. It is expected that in such networks, all distributed energy producers and consuming entities will be highly interconnected via information flows, many of which will depend on M2M interactions (Karnouskos & Terzidis 2007). In that sense, a paradigm changes from existing passive and information-poor to active information-rich energy networks, which reverses the trend of one-way flow because the electricity networks that were initiated is underway. Such a future infrastructure is expected to be service-oriented and give rise to new innovative applications that will drastically change our everyday environment. The bidirectional information exchange will put the basis for cooperation among the different entities (Karnouskos 2011b), as they will be able to access and correlate information that up to now either was only available in a limited fashion (and thus unusable on a large scale) or extremely costly to integrate. Examples of such emerging capabilities are demand-side management and local energy trading (Karnouskos 2011b, US DOE 2006).

M2M-capable assets that are increasingly introduced not only in the industrial domain such as sensors, SCADA/DCS systems, etc., but also on the end-users, for example, as Electric Vehicles (EV), white label appliances, and even light bulbs, lie at the heart of the Smart Grid. One example is GreenWave Reality's WiFi-aware light bulbs, which can be controlled by other devices such as smartphones (GreenWave Reality 2013). The capability of offering fine-grained information about a device's energy signature as well as control capabilities depicts a shift towards a highly observable and controllable infrastructure that can dynamically adjust to the real-world conditions. The latter empowers the Smart Grid vision with massive monitoring and management capabilities that were not feasible before. In addition, the fine-grained information provided is also used for high-performance analytics, which benefits a variety of existing approaches ranging from technical monitoring, to predictive maintenance and accurate economic model assessment during the lifecycle of the assets.

Apart from the traditional M2M aspects tackled in the bulk generation and distribution domains, of particular interest to M2M is also the customer, operations, markets, and service provider domains, especially those that directly or indirectly are affected by a previously uncontrollable and passive infrastructure (i.e. that of end-users). Hence, we will take a closer look at how M2M empowers end-users and what new capabilities they can exercise in Smart Houses and Smart Grid Cities.

12.2 Smart metering

Traditional utility management processes today involves the collection of metering data from a centralized point (the meter) in the household once or twice a year. With the emergence of smart metering, the collection of data from households has increased, with the aim to have measurements (or less) at 15-minute intervals. This coupling of smart meters to an infrastructure that can monitor at such a fine-grained form the grid consumption for residential households has profound implications if coupled with other tools such as data analytics (covered in Sections 5.6 and 5.7).

Many utility providers today strive towards offering new energy services that will enable customers to get a better grasp of their energy consumption by seeing the amount of energy they consume in short intervals (the aim is to be near real-time or even down to minute resolution). Coupled with variable tariffs, this may provide customers with a view on the upcoming costs and assist end-users in adjusting their energy consumption behavior (which may lead to more sustainable energy management) and avoid cases such as "bill shock" (where the users get shocked after 1−2 months when they discover the high electricity costs). To realize such services, key IoT aspects such as integration, remote monitoring, and remote management are needed.

Figure 12.3 depicts the paradigm change from an infrastructure delivering metering data for billing purposes towards a general-purpose monitoring infrastructure that acts as an enabler for a multitude of stakeholders and value-added services. Data generated include not only the energy metering data such as consumption (or production of energy), but also other data (e.g. related to power quality, device status, etc.), which may provide additional added value towards asset management. Such smart metering infrastructures are currently under investigation for their performance (Karnouskos et al. 2011a), scalability, as well as the value-added information they can deliver beyond smart metering (e.g. towards energy management, asset management, integration of heterogeneous systems, etc.) (Karnouskos et al. 2010b).

The new smart meters, or their extensions (such as dedicated devices depicting energy consumption and other info), are used as communication media where the utilities push information (for example) about the current

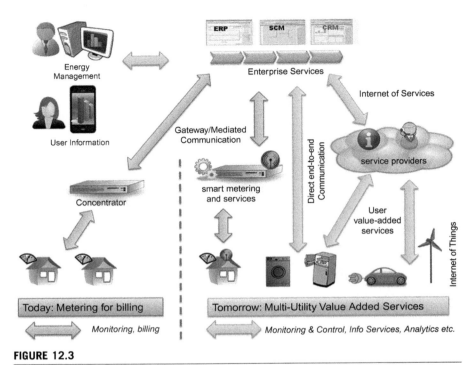

FIGURE 12.3

Smart Metering enabling multi-utility value-added services.

energy tariff, upcoming maintenance, costs, etc. By additionally adopting varying tariffs and projecting that to the end-users, the aim is to have soft control of the infrastructure; more precisely, the expectation is that due to higher-prices at key times during the day, the users will shift part of their activities to non-critical times. When the latter is realized successfully from a large number of users (critical mass), it can have a significant impact on the infrastructure resources and help tackle aspects such as peaks of consumption, which would be too costly. On a similar train of thought, lowering the price implies shifting energy consumption to those intervals where the energy is available (usually coming from intermittent resources such as wind parks, photovoltaic (PV) farms, etc.), but not enough consumption is present, for example. Although the industrial infrastructures have more fine-grained control, this is the first step towards introducing control in residential households (which were previously passive, unmanageable consumers of energy only).

M2M is key to this bidirectional communication between the utilities and their customers. The millions of smart meters required to realize the vision of fine-grained monitoring (and potentially control) are currently under assessment and deployment worldwide. Indicatively, it is expected that around 250 million meters will be deployed in Europe, with around 80% of them (ca. 200 million) by 2020 at a cost of approximately €30 billion (Giordano et al. 2013). Smart metering customer behavior trials indicate that there might be benefits for all involved stakeholders, including reduction of electricity bills, better usage of electricity, etc., but more importantly, other benefits may also exist, such as improvement of competition in retail markets, better demand—response, and future coupling with energy management and automation systems.

12.3 Smart house

We are witnessing an increased penetration of networked embedded systems in modern appliances, which transforms the residential infrastructure to an M2M-enabled one, and connects it to the Smart Grid. Hence, white goods such as refrigerators, washing machines, microwaves, etc. are no longer passive black boxes that consume energy, but intelligent devices whose behavior may be actively communicated (monitored) via networking interfaces, as well as adjusted (controlled) via interactions with other systems (e.g. an energy management system). All these networked energy-consuming and/or -producing devices (generally called prosumer devices), exponentially expand the M2M infrastructure related to energy, with billions of devices that can now be active participants in energy management efforts (Figure 12.4).

A typical example of understanding the significance of the inclusion of M2M residential infrastructures is the following. During the summer, many countries face electricity shortages that may lead to a blackout. Up to now, many such events were tackled with manual processes, i.e. the grid operator (in parallel to increasing the energy production) may contact heavyweight energy consumers and ask them to reduce their load. However, blackouts still occur and are costly for all stakeholders. Clearly this can be done with heavy industrial partners, however, such an approach doesn't scale and also has other drawbacks. On the other hand, if peaks coming from the residential consumers could be reduced (or shifted), that would greatly ease energy management and make it more efficient. Hence, in the above scenario, instead of contacting only a limited number of industrial consumers, now millions of residential consumers could be contacted and assist by adjusting proactively their energy consumption.

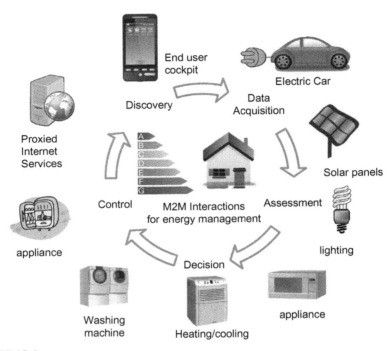

FIGURE 12.4

M2M interactions among various devices (P2P) and with an energy management system in a Smart House.

For instance, all unnecessary devices could be turned off momentarily in case of urgent needs. Prioritization on the device level (e.g. don't turn off the fridge or reschedule the washing machine, etc.) as well as on process or location (e.g. don't cut off power in hospitals and emergency infrastructures) could be applied. In any case, the key message here is that similar results can now be achieved on a large scale if a critical mass of residential consumers can be reached.

Several research projects (Giordano et al. 2013) focus on the integration of the Smart House and its appliances to the Smart Grid, and investigate how energy could be better managed in order to increase efficiency, without noticeably impacting quality of life. One example of such a project is SmartHouse/SmartGrid (Karnouskos et al. 2010b, 2010c), which focused on the in-house as well as the Smart House with enterprise systems integration and trialed various approaches, as depicted in Figure 12.5. It is unlikely that a "one-size-fits-all approach" will work in such cases, but rather many and diverse approaches for managing

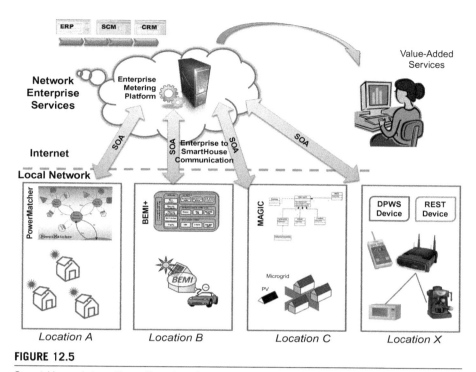

FIGURE 12.5

Smart House integration with enterprise services, as demonstrated by the SmartHouse/SmartGrid project.

residential areas and connecting Smart Houses and devices to the Smart Grid, systems integration will be a key issue for ensuring that benefits from such solutions are delivered. SmartHouse/SmartGrids build on the following elements: (1) in-house energy management based on user feedback, real-time tariffs, intelligent control of appliances, and provision of (technical and commercial) services to grid operators and energy suppliers; (2) aggregation software architecture based on agent technology for service delivery by clusters of Smart Houses to wholesale market parties and grid operators; and (3) usage of Service Oriented Architecture (SOA) (Karnouskos et al. 2011a) and strong bidirectional coupling with the enterprise systems for system-level coordination goals and handling of real-time tariff metering data.

Within the household, appliances and devices are integrated via some form of gateway or concentrator that connects to the Smart Grid (as depicted in Figure 12.5). Several integration approaches were experimented with, e.g. the PowerMatcher, the bi-directional energy management interface (BEMI), the Magic system, as well as direct web service integration (Warmer et al. 2009). These represent not only different technical integration approaches, but also different models of management. The integration via DPWS/REST enabled the direct integration of SOA-ready devices (i.e. devices hosting native web services) or at gateway level with enterprise systems, easing their management. Similarly, in the mediated interactions, various approaches such as multi-agent systems or middleware were used, which offered delegation of the intelligence and decision-making near to the actual infrastructure (the devices).

What is also required for viable business cases with regards to Smart Houses as part of the Smart Grid is the integration of in-house services with enterprise-level services. The last includes typical business-to-customer (B2C) services such as billing, but also other business-to-business (B2B) services such as the interaction among different players such as the distributed generation (DG) operator, energy retailer, wholesale market, and others.

12.4 Smart energy city

With an increasing amount of the grid core infrastructure embracing IT technologies, the Smart Grid is advancing and is being complemented with similar functionality coming from Smart Houses. The same principles apply increasingly to other parts of the modern city energy infrastructure, including buildings, traffic systems, renewable energy parks with PV/wind turbines, the public lighting system, etc., just to name a few. Hence, we witness the metamorphosis of cities into smart energy cities where innovative approaches towards energy efficiency can be applied at a new level and enable their better energy footprint management.

Future smart cities are expected to provide superior quality of life to its citizens, and as we see in Chapter 14, M2M and IoT will greatly boost such efforts. However, in a similar way, new innovative services and applications are expected to empower better understanding and tackling of energy-related issues. For instance, monitoring applications could offer a real-time view on a city's key performance indicators, such as CO_2 footprint, energy consumption increase, penetration of renewables, etc. For example, citizens may now be able to calculate with smart apps the environmental impact of their in-city traveling options. Alternatively, public authorities may be able to better identify energy-hungry processes at a citywide level, and plan how to tackle those as well as a more sustainable city expansion. The SmartKYE project (2013) is currently developing such

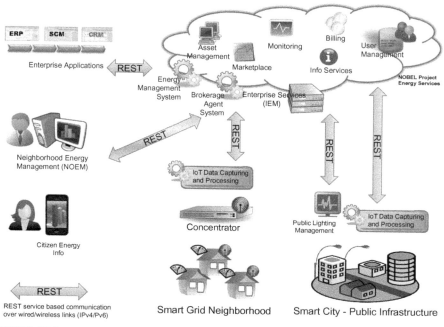

FIGURE 12.6

Smart City Energy Services.

an energy cockpit for smart cities that can help decision makers to make more informed decisions in the future (Figure 12.6).

Monitoring and control of the M2M infrastructure plays a pivotal role in order to not only extract the necessary information for understanding key energy processes, but also to be able to apply control when decisions are taken. Once this is a reality on a large scale, new innovative applications that depend on M2M data as well as their control capabilities can be realized, with impact on individual citizens as well as citywide. The NOBEL project (2013) dealt with integrating information coming from various energy aspects of the city infrastructure such as energy producing/consuming customers (e.g. residential houses, buildings, etc.) as well as the public lighting infrastructure. By tapping into the extended information of smart metering, it was possible to enhance existing services, e.g. real-time monitoring of energy consumption, real-time billing, asset management, etc. (Karnouskos et al. 2013b); it was also possible to provide new innovative services such as energy trading (Ilic et al. 2012a) in a citywide energy marketplace, direct interaction between the energy provider and the

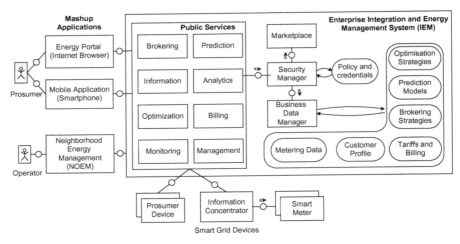

FIGURE 12.7

Enterprise IEM System used in the NOBEL project trial.

consumer, and identification of failures (Ilic et al. 2012b) in Smart Grid infrastructures, etc.

To be able to provide value-added energy services, a platform was created (Karnouskos et al. 2013b) which eased the M2M interactions over IPv4 and IPv6 (Höglund 2012), and especially data collection, which was then evaluated and provided to the respective services. The architecture of the platform is depicted in Figure 12.7. The platform (Integration and Energy Management System, IEM), as well as the applications depending on it, have been extensively tested and used operationally in the second half of 2012 as part of the NOBEL pilot project that took place in the city of Alginet in Spain. Data in 15-minute resolution, of approximately 5,000 meters, were streamed over a period of several months to the IEM, while the IEM services were making available several functionalities ranging from traditional energy monitoring up to futuristic energy trading (Karnouskos et al. 2013a, 2013b).

More specifically, as depicted in Figure 12.7, we can distinguish the following key functionalities commonly offered via mobile (e.g. smartphones, tablets, etc.) and traditional (e.g. desktops, web browsers) devices, other more complex services, and end-user applications:

- **Energy Monitoring**: For acquiring and delivering data related to the energy consumption and/or production of prosumer devices.
- **Energy Prediction**: For forecasting consumption and production based on historical data acquired by IEM and other third party services (e.g. weather data).

- **Management**: For handling the asset, user, and configuration issues in the infrastructure.
- **Energy Optimization**: For interacting with existing assets of the Smart City; as a proof of concept the public lighting system was used for energy balancing.
- **Brokerage**: Offering energy trading to all prosumer citizens who in a stock-exchange manner could interact via the platform and buy potentially cheaper energy or sell excess production from their PV panels.
- **Billing**: Offering a real-time view of the energy costs and benefits (from transactions on the Smart City energy market), hence avoiding "bill-shock" scenarios.
- **Other**: Value-added services offering bidirectional interaction between the users and the energy provider, such as notification for extraordinary events, etc.

The hands-on experiences (Karnouskos et al. 2013a, 2013b) in 2012 within the NOBEL project revealed several benefits for Smart Grid Cities, as well as several aspects that are crucial when dealing with M2M infrastructures. Designing open services is a must, and the needs of various stakeholders need to be considered. With the enhanced monitoring capabilities at the IoT layer, as well as the increased access and correlation to enterprise data, the era of IoT "Big Data" reaches the Smart Cities. This has profound implications on the apps and services, depending not only on the processed outcome of the data, but also on their quality and timely acquisition. To this end, e.g. validation of data values and syntax based on model semantics, correct time-stamping, duplicate detection, security validation and risk analysis, anonymization, data normalization, estimation of missing data, conversion to other formats or models, etc., may be necessary prior to release of the data for further processing or consumption (Karnouskos et al. 2013b).

12.5 Conclusions

M2M and the Smart Grid are a very powerful combination. M2M has the capability of revolutionizing the core of the electrical grid infrastructure and transforming it to a truly smart one. In combination with the IoS and the M2M, new applications and services can be realized, effectively tackling older problems and providing innovative solutions. The ongoing efforts in smart metering are only a small part of what the future energy infrastructure will be able to do, and we have already shown that based on this data, various stakeholders can enjoy benefits at the Smart House as well as at the Smart City level. To harness the benefits, there are challenges both technological and social that need to be addressed. The future Smart Cities will need to be built on principles of cooperation, openness/interoperability, and trust. Extracting and understanding the business-relevant information under temporal constraints, and being able to effectively integrate it in solutions that utilize the monitor–analyze–decide–manage approach for a multitude of domains, is challenging (Karnouskos et al. 2013b).

Commercial Building Automation

13.1 Introduction

A Building Automation System (BAS) is a computerized, intelligent system that controls and measures lighting, climate, security, and other mechanical and electrical systems in a building. The purpose of a BAS is typically to reduce energy and maintenance costs, as well as to increase control, comfort, reliability, and ease of use for maintenance staff and tenants.

Some example use cases:

- Control of heating, cooling, and ventilation based on time of day, outside temperature, and occupancy (e.g. Morning Warm-up).
- Automatic control of air handlers to optimize mix of outside air in ventilation based on, for example, inside temperature, pressure, and time of day.
- Supervisory control and monitoring to allow maintenance staff to quickly detect problems and perform adjustments.
- Outsourcing of monitoring and operations to a remote operations center.

From Machine-to-Machine to the Internet of Things: Introduction to a New Age of Intelligence.
DOI: http://dx.doi.org/10.1016/B978-0-12-407684-6.00013-9

- Data collection to provide statistics and facilitate efficiency improvements.
- Alarms for high carbon monoxide and carbon dioxide levels.
- Individual metering per apartment (to give incentive to save energy in multi-tenant buildings).
- Intrusion and fire detection.
- Building access control.

A BAS is normally distributed by nature to allow every sub-system to continue operation in case of failure in another system.

A BAS consists of the following components (Figure 13.1):

- Sensors (i.e. devices that measure, such as thermometers, motion sensors, and air pressure sensors).
- Actuators (i.e. controllable devices, such as power switches, thermostats, and valves).
- Programmable logic controllers (PLCs) that can handle multiple inputs and outputs in real time and perform regulating functions, for example.
- A server which monitors and automatically adjusts the parameters of the system, while allowing an operator to observe and perform supervisory control.
- One or more network buses (e.g. KNX, LonWorks, or BACnet).

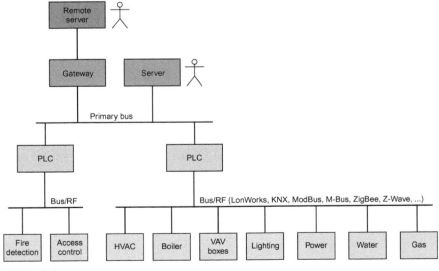

FIGURE 13.1

Central parts in a BAS.

We have divided the case study into two phases. In Phase One we give an example of what is commonly available today in regards to building automation. In Phase Two we explore new opportunities for building automation, such as the Smart Grid and the IoT.

13.2 Case study: phase one — commercial building automation today

13.2.1 Background

Company A wants to improve energy efficiency in their buildings and become GreenBuilding Partner (2013) certified, which requires lowering their energy consumption by at least 25%.

After discussions with a building automation company (Company B), they have come to understand that this is a very good investment that will quickly justify itself in terms of reduced energy costs. They agree on a five-step plan that starts with collecting data from the buildings, followed by analysis, adjustments, and connecting the systems in the buildings to a local server, and finally connecting the buildings to a remote operations center.

They can now start with collecting data from existing systems. In some cases this requires new meters to be installed. Everything from water usage to heat and electricity consumption is logged continuously, as well as performance of the ventilation and room temperatures.

By comparing the key performance indicators with comparative figures, the need for corrective actions is assessed and used as a basis for an action plan that consists of adjusting the existing systems and installing new software. These adjustments quickly increase the efficiency of the systems and are continuously optimized during the project. Examples of adjustments are hot water temperature, improved control of indoor temperatures, as well as better control of fan and pump operation to avoid unnecessary operation.

One of the most central features of the improved system is the new web-based E-report. It provides information about current energy consumption and other key parameters from the buildings. This information is used to make both short-term decisions as well as long-term planning. Everyone has access to the web portal because it's not only important for the maintenance staff, but also needed to create awareness across everyone in the company.

The next phase of the project consists of connecting the systems in the buildings and analyzing the dynamics to be able to perform intelligent control. This both improves performance as well as reduces maintenance costs.

The final step to completion involves setting up a web-based Supervisory Control and Data Acquisition (SCADA) system for remote monitoring of the building systems. Through the web portal, the users can access information from the buildings in a coherent manner. Company A decided to outsource the operations and daily maintenance of the systems to Company B by utilizing their cloud-based offering. Company B's remote operations center is continuously monitoring the building systems. When building system operations deviate from their expected behavior, Company A's maintenance staff and their supervisors are notified by SMS and email. Typical events that can trigger a notification are, for example, mechanical failures or undesirable temperature deviations. Apart from notifications, Company B can also assist with equipment operation and adjustments remotely. For Company A, this arrangement is perfect because their in-house maintenance staff can respond to an alert 24 hours a day.

For Company A, the most important improvement has been the 35% energy reduction after the completion of the project. Another critical aspect has been the knowledge transfer from the experts at Company B that allows Company A to maintain the efficiency of the systems as well as the ability to continuously improve the operations of them.

13.2.2 Technology overview

Figure 13.2 depicts the setup for Company A.

Each building is equipped with a set of meters and sensors to measure temperature, water consumption, and power consumption, as well as one or more PLCs.

As seen in Figure 13.2, the PLCs perform real-time monitoring and control of the devices in the building. They also feature a user interface for configuration and calibration of (for example) the regulators, curves, and time relays. It is possible to remotely configure the PLCs from the Operations Center using the PLC Control system, which is connected to the PLC via a 3G-modem and an Internet Protocol (IP) modem that converts between RS-485 networks and Transmission Control Protocol/Internet Protocol (TCP/IP) networks. The PLCs communicate with the devices using several protocols, such as M-BUS, analog, digital, and Z-Wave, which is a low-power radio mesh-network technology. All logic necessary to operate the buildings is contained within the PLCs, allowing for minimal bandwidth requirements on the connection towards the

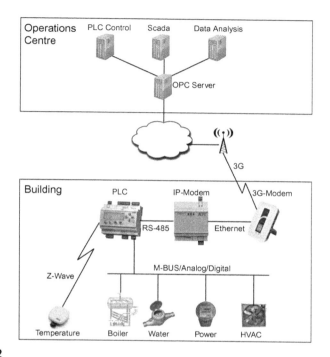

FIGURE 13.2

Illustration of the BAS.

Operations Center as well. It also means that the building systems can remain fully operational during periods of network outage.

The OLE for Process Control (OPC) server provides access to data, alarms, and statistics from the PLCs. When a value is requested from a user, a request is sent from the user's OPC Client to the OPC Server, containing an OPC Tag that identifies which PLC to contact and which value to ask for. The type of OPC communication used is called OPC Data Access. The OPC server then contacts the PLC in question and asks for the value using a protocol supported by the PLC (LonWorks or ModBus).

The SCADA system is used for operational monitoring of the buildings and provides information from all the relevant building systems. It uses the open and standardized OPC protocol, which enables integration with devices from many different vendors. The maintenance and operations staff can connect to the system using a web browser with a username and password to access dynamic flowcharts, drawing tools, timers, set points, actual values, historic readings, alarm management, event logs, as well as configuration for notifications over email, fax, or SMS.

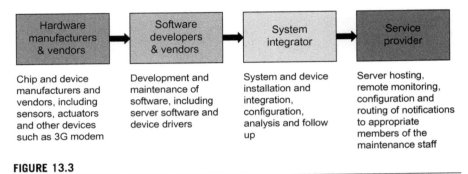

FIGURE 13.3

Applied value chain for Company A's system.

The Data Analysis server logs all historical readings from the buildings and makes it possible to follow up on different aspects of the energy and resource consumption, satisfying the varying needs of the tenants, economy department, and landlord. Through the OPC server it's possible to gather readings from all the building systems, regardless of vendor. Typical reports include trends, cost, budget, prognosis, environment, and consumption of electricity, heating, water, and cooling.

13.2.3 Value chain

Figure 13.3 shows an applied value chain.

13.3 Case study: phase two — commercial building automation in the future

13.3.1 Evolution of commercial building automation

Two major factors will drive the evolution of Building Automation: information and legislation (Figure 13.4).

Access to well-packaged information will provide the basis needed for decisions and behavioral changes. This can (for example) be electricity prices or where and when energy is used, and will allow for well-founded decisions that provide the best results.

Legislation, and taxes or tax credits to some degree, will provide the second driver. Legislative demands on green buildings and the Smart Grid will give rise to new opportunities, such as Demand/Response, Micro Generation, and Time-of-Day Metering.

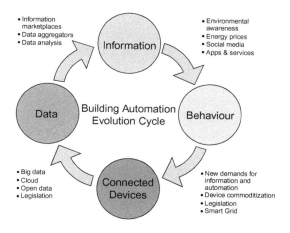

FIGURE 13.4

Building automation evolution cycle.

Market growth will result in economies of scale, standardization and commoditization, driving down prices, and increasing availability of devices and services. It will be possible to buy advanced devices off-the-shelf, perform installation, and connect them directly to service providers on the Internet.

13.3.2 Background

A few years have passed and Company A has decided to outsource the maintenance of its buildings to a local contractor who provides services to several other customers in the neighborhood. This will save money since that will enable them to utilize a shared caretaker pool.

At the same time they plan to upgrade their buildings to become fully automated with, for example, occupancy sensors, automated lighting, and integrated access control. To make this cost efficiently, they intend to make use of the existing IP infrastructure in the buildings, which also saves on operating expenditures as the network administrators can also manage the BAS infrastructure. According to studies, a converged IP and BAS network can reduce maintenance costs by around 30% while also lowering the initial investment for installation and integration by around 20% (according to studies performed by Cisco®). A shared infrastructure also leads to increased energy efficiency.

New political incentives in regards to energy efficiency have increased the development pace in the building automation area. Many neighboring buildings in Company A's area are now fitted with building automation, which allows for sharing of information and resources. The increased customer base has also enabled new niches in the value chain, which has been split up to a large degree. Where before the rule was to have one single integrator and service provider, we now see a multitude of new actors, such as specialized service providers for remote monitoring, security, optimization, data collection, and data analytics. This allows Company A to choose freely what combination of service providers to use, while also providing a smooth transition when moving to a new provider. This is made possible by a new niche in the value chain: the Cloud Service Broker (Figure 13.5).

The process of integrating with the maintenance contractor's systems is simplified by the service broker because it provides immediate access to Company A's BAS. The caretakers can use their own specialized software as the service broker provides a bridge that can convert between several common protocols used for building automation.

When it comes to selection of devices, Company A opts for using standardized protocols to avoid vendor lock-in. They also decide to keep certain parts of the old system, as these would be too costly to replace. To still benefit from a fully integrated system, they also invest in a constrained application protocol (CoAP) gateway that translates between legacy devices and the new system.

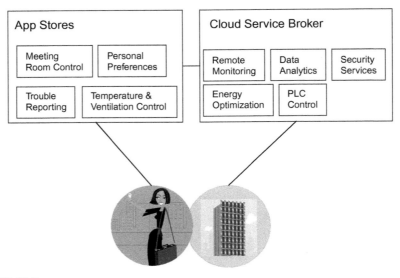

FIGURE 13.5

Cloud Service Broker.

By exporting historical data and configuration parameters from the old OPC Server it's possible for Company A to choose new service providers that can replace the old systems for PLC control, SCADA, and data analytics with a minimum of manual effort.

As an added service, the new platform also provides data brokering. This gives access to a multitude of data sources, such as the following:

- Historical and current KPIs to similar buildings.
- Integration with local government facilities.
- Weather forecast information.
- Utility prices, both current and future.

Apart from providing access to new service providers, the cloud broker also hosts a client API that enables third-party app developers to create smartphone applications. A number of users have purchased apps that allow them to do the following, for example:

- Control HVAC settings in meeting rooms.
- Report problems and service requests.
- Integrate with Outlook to adjust meeting rooms in advance.
- Create personal profiles to automatically adjust room settings.
- View instant and historical personal energy consumption and compare to others using social media.

13.3.3 Technology overview

Thanks to the rapid development of technologies for IP Smart Objects, it's now possible to use IP for both constrained devices, such as battery-powered sensors and actuators.

The new system is to a large degree based on IP technology (Figure 13.6). There are several IP-based protocols to select from, but in this case CoAP and Sensor Markup Language (SenML) were selected. CoAP provides both automatic discovery as well as a semantic description of the services the device provides. This drastically reduces installation costs, as much less configuration is needed. CoAP is similar to Hypertext Transfer Protocol (HTTP), but is binary to reduce the size of the messages. It also defines a Representational State Transfer (REST)-like Application Programming Interface (API) optimized for M2M applications. As with HTTP, a format for the content is also needed, in this case SenML, which is used as a format for sensor measurements and device parameters.

As mentioned before, there are still a few legacy devices, and these need a gateway to enable communication with the IP-based systems.

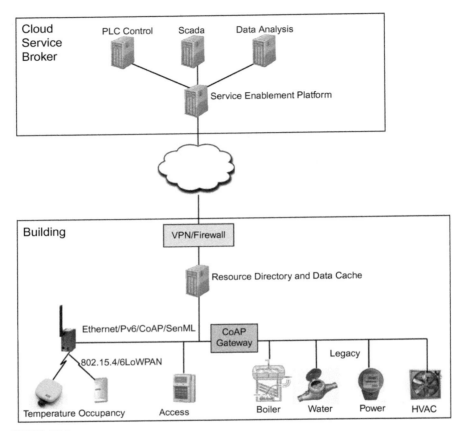

FIGURE 13.6

Architectural overview of the upgraded system.

A local Resource Directory and data cache is also installed to keep track of all the devices in the company network. This allows local lookups of devices and data, and serves as a safeguard in case of failure.

To protect the system from intruders, a normal network firewall is used. For the connection towards the service broker and the service providers, a permanent Virtual Private Network (VPN) connection is established.

Historical data is exported from the OPC server to the Cloud service broker's data storage to make it available to new service providers. Apart from data storage, the Cloud service broker also provides management functionality for control and data access, as well as access to specific service provider's management portals. It also offers a global Resource

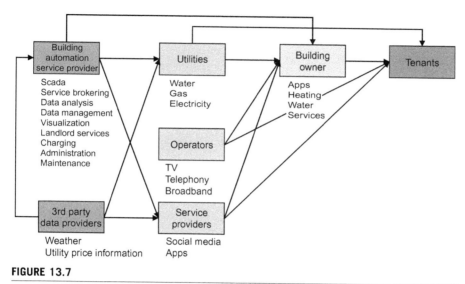

FIGURE 13.7

Evolved value chain for building automation.

Directory with semantic resource and data description, along with a contextual model that covers schematics, geospatial information, and indoor location.

13.3.4 Evolved value chain for commercial building automation

As the demand for M2M services grows, new niches in the value chain will emerge, such as information brokers, service brokers, and service enablement providers (Figure 13.7). These will enable new use case areas by allowing vertical domains to be integrated (e.g. security, energy, waste management, police, and public transport). It will also provide the openness needed for third-party service providers to create apps and social media integration, to allow for (for example) comparisons with neighboring buildings, competitions, and end user involvement. Privacy and security will, however, be essential to build up the trust needed for this ecosystem to develop.

In addition to new use case areas, major improvements can be expected within the areas of efficiency, convenience, comfort, reliability, and safety.

Smart Cities

14.1 Introduction

We are now a "city planet" (Brand, 2006). As of 2007, 50% of the world's population was living in cities rather than rural areas (UN-HABITAT, 2011). Moreover, cities will continue to grow—it is predicted that 70% of the world's population will be living in cities by 2050. In addition to this fundamental shift in the organization of human society, we are also faced with increasing natural resource constraints, marked increases in population, and a restructuring of the global economy. Existing cities and the new ones that will be built will therefore need to handle massive tensions in three spheres simultaneously: environmental impact, economic growth, and social evolution. This will affect every city in the world; for example, failure to provide new employment opportunities in the face of massive economic change and provide adequate environmental protection will have a profound effect on the inhabitants of cities. Competition between cities is also set to increase, as cities need to create post-industrial identities that are attractive to employees and employers, and tourists, and can also provide communities the ability to combine their resources to overcome common problems rather than relying on increasingly scarce government resources. The pressure and tensions experienced within cities today are only just beginning.

From Machine-to-Machine to the Internet of Things: Introduction to a New Age of Intelligence.
DOI: http://dx.doi.org/10.1016/B978-0-12-407684-6.00014-0

As a result, so-called "Smart Cities" have rapidly become a hot topic within technology communities, and promise both improved delivery of services to end users and reduced environmental impact in an era of unprecedented urbanization. Both large, high-tech companies and grass-roots citizen-led initiatives have begun exploring the potential of these technologies. A multitude of so-called "smart city" projects are being developed across the world. This chapter covers some of the aspects involved related to M2M and IoT and presents a few example use cases.

While similar sorts of concepts have been around for several decades, appearing as early as the 1980s (Batty, 2012), the recent advent of smart-phones and cheaper sensor technology means that digitally enabled, or "smart," cities are fast becoming a real-world possibility (IMTNEC, 2011).

14.2 Smart cities—the need

The need for smart cities has been emerging for some time, in particular due to the increasing urbanization during the last 50 years. Cities now account for 75% of the world's greenhouse emissions, but take only 2% of the earth's surface area. At the same time, the size and economic outputs of cities is on par with small nations—only 600 urban centers generate about 60% of global GDP. For example, Tokyo ranks as one of the world's top 15 economies with 35 million people and nearly $1.2 trillion in economic output—this is larger than either India or Mexico.

Cities need to provide for the majority of the world's citizens while rapidly decreasing environmental impact. As a result, the need for smart cities is discussed in a variety of settings, including the need for climate change mitigation to providing cities that are attractive to citizens not just economically but also socially—providing places that are enjoyable to live in. In addition to this, cities need to spend a large amount of money on urban infrastructure—$350 trillion or seven times the current global GDP in the next 30 years (information marketplaces).

Technical innovation has the opportunity to provide some of the solutions to the era of urbanization, thus helping to both reduce environmental impact and also create new jobs and promote growth and citizen engagement in the democratic process. As we discussed in Chapters 2 and 3, technology advancement is providing the opportunity to manage and handle broad social issues at a much deeper level than previously. This is thanks to the processing power and storage capacity of computer technology coupled with mobile broadband, which delivers computing capacity directly into the hands of citizens. Information and Communications

Technology (ICT) can be applied in the built environment to help solve issues such as traffic congestion, energy consumption, and behavioral change. In addition, however, the positive externalities produced by ICT can help create new jobs, innovation, and economic growth. The technology-enabled city is an untapped source of sustainable growth. By unlocking technology, infrastructure, and public data, cities can open up new value chains that spawn innovative applications and information products that make possible sustainable modes of city living and working (information marketplaces).

In order to become "smart," however, cities do not just face technical challenges, they face organizational and educational challenges. City leaders need to understand how to create integrated technology strategies, operating frameworks, and incentives for future digital cities.

14.3 Smart cities—a working definition

As discussed above, smart cities are a complex and multi-layered problem with multiple stakeholders that go far beyond pure technology solutions. For the purposes of this chapter, we therefore define a smart city as one that uses data and ICT in order to:

- Manage and optimize existing infrastructure investments and plan for new investments more effectively.
- Provide more efficient, new, or enhanced services to citizens.
- Reduce organizational silos in cities' service delivery and create new levels of cross-sector collaboration.
- Assist city and government's progress toward meeting climate change mitigation and adaptation goals.
- Enable innovative business models for public and private sector service provision.

By aligning the interests of stakeholders and employing new technologies and new market mechanisms, cities will be better able to capture the full value of ICT investments made in order to become a smart city.

14.4 Smart cities—some examples

There are many different types of "smart city" solutions; these include:

- **Smart Transportation** solutions for parking, traffic monitoring, public transport, or municipal fleet management.

- **Smart Healthcare**, which includes electronic records management, hospital asset management, and remote monitoring systems.
- **Smart Education**, including eLearning, MOOCs, and connected campuses.
- **Security and Public Safety**, such as video surveillance, enhanced analytics, and workflows for emergency services.
- **Smart Buildings**, which include smart meters, light, heating, internal security systems, and water and waste management.

14.5 Roles, actors, engagement

One of the biggest complexities associated with cities is the sheer number of actors involved in them. Moreover, many of these actors have multiple roles within a city context. In comparison to the other use cases in this book, therefore, smart cities are much more difficult to implement due to the sheer number of actors and stakeholders. In fact, smart city solutions are often the combination of several of the other use cases described—for example, a combination of a participatory sensing and a smart grid scenario. Figure 14.1 illustrates some of the complexity within cities. A city may be viewed as a nexus of several infrastructures and social constructs. For example, cities have buildings not just for corporations, but for citizens and government services as well. Infrastructure covers a broad range of meanings, from the actual pipes laid down for transmission of water or the cables for transmission of electricity, all the way through to the garbage collection and recycling mechanisms.

In contrast to the other use cases in this book, people are both key barriers and key opportunities for change within a smart city use case. A citizen, for example, holds a number of different roles within a city; for example, as a parent who needs to interact with schools in the district, an employee who needs to access city infrastructure in order to get to work, a voter, or as an end user of city services such as waste management.

In addition to the complexity of the number of roles that a citizen plays within a city context, the infrastructure itself is deeply embedded in a number of places—from the urban spaces that form the city streets to the subterranean networks of infrastructure for utilities such as electricity, water, and waste that lie underneath the streets of every major city in the world. Moreover, a city cannot in today's world survive without its hinterland—the surrounding spaces and other towns and cities that help deliver the natural resources and other services that it uses. As a result, one of the main issues

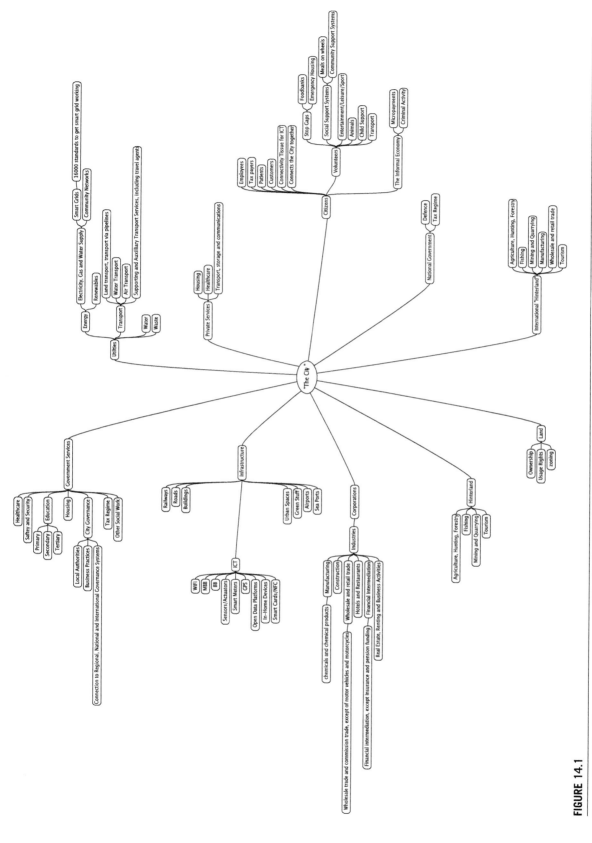

FIGURE 14.1

Complexity within cities.

that cities face today is the effective management of multiple interacting logistics issues—bringing food and water into a city, removing waste, bringing in and delivering goods and services to companies and citizens alike, as well as enabling the same citizens and companies to send their goods and services to other cities.

We cover in the following section a detailed use case on transport and logistics, the role that M2M/IoT will play, and the use of data associated with different actors in the creation of an information marketplace.

14.6 Transport and logistics—an IoT perspective

"Transport" generally corresponds to the delivery of people and goods via road, rail, air, and sea (Classifications 49, 50, 51, 53 in the United Nations Statistics Division's *International Standard Industrial Classification of All Economic Activities*) (United Nations Statistics Division 2008). "Transport infrastructure and services" refers to all infrastructure and services that support transport (Classification 52), as well as the manufacture of vehicles and equipment for monitoring and servicing transport.

These sectors are those that deliver day-to-day travel and delivery services in and around cities for home, work, and leisure activities. For the sake of simplicity, these are collectively called "intraurban" transport to distinguish it from long distance intercity transport.

Cities face dramatic challenges in the coming decades with the management of transport systems. It is predicted that soon there will be nearly 6 billion cars on the planet; in addition to this, 180,000 people move into cities each day, placing pressure on already stretched transport infrastructures.

In the following example, we investigate the use of IoT to create new solutions to solving these issues. Due to the complexity of the transport sector in cities, we apply the framework outlined in Chapter 3 for information marketplaces. Firstly, however, we need to have a basic understanding of the physical infrastructure that we are dealing with.

14.6.1 Physical infrastructure for transport

Transport infrastructure is, as mentioned previously, complex and deeply embedded in the natural environment of a city. From roads to rail networks, the decisions made about how to make decisions about when and where to implement transport infrastructure are critical to the future success of a city. Today's modern transport infrastructure industry has often been split up into several areas of control; while there are national differences between how transport is implemented, there are generally some broad categories. Each of these actors may implement IoT in a different way:

Manufacturers: Manufacturers build and sell infrastructure assets (e.g. trains, tracks, buses). Example companies include Siemens.

Infrastructure Managers: Infrastructure managers buy infrastructure from manufacturers and implement, manage, and maintain the infrastructure on behalf of a city or a government. Examples include the rail networks in the UK or the road infrastructure authorities responsible for building and maintaining national road infrastructure.

Operators: Operators are those companies that run the actual services on the infrastructure. These companies often set timetables and are the ones that most end-user consumers know and have customer relationships with. Examples include Virgin Trains in the UK rail industry, or Greyhound coaches, who operate coaches on the road infrastructure in the U.S.

Maintainers: Maintainers are those entities that are responsible for the day-to-day running of different transport infrastructures. For example, ensuring that the trains are kept serviceable and that roads are properly maintained. In many countries, these are outsourced operations paid for by the Operators.

End Users: End users are the people who are actually using the services; they can be train passengers, coach passengers, and in some cases, even drivers of their own cars.

Regulators: A key aspect of the transport infrastructure in any nation is the regulator. These provide regulations that companies need to abide by in a variety of situations. Examples include anti-monopoly regulations, anti-price setting, management of timetabling information, and also setting health and safety standards for end users and staff working on the infrastructure itself.

The actors involved in the physical infrastructure are illustrated in Figure 14.2:

Each of these actors will have a different requirement from IoT and M2M installations. These are illustrated in Table 14.1.

M2M solutions have therefore enabled a large amount of information to be gathered across the transport infrastructures of different nations. These information sources from sensors, etc., however, are currently held in data silos—the actors within the value chain do not share information in

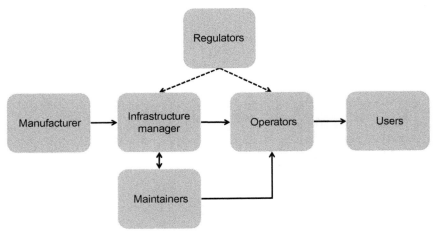

FIGURE 14.2

Actors in the Transport Value Chain.

Table 14.1 Actor Requirements from IoT and M2M Installations		
Actor	**Information**	**Example Data Sources**
Manufacturer	• Where is my equipment? • How is it being used? • How is it performing? • How is it being maintained? • Does it need redesign? Data feeds for redesign. • What are the new markets for my product? • What needs replacing and where? • What are the levels of utilization it is seeing? • What other systems is it interacting with (that I don't know about)? • What is the present state of the equipment (accidents)? • Management of recalls, safety notices.	• Sensors and aggregated data readings. • Customer relationship databases.
Infrastructure Managers	• State of Infrastructure. • How are maintainers performing?	• Sensors on tracks. • Switches (rail). • Maintenance schedules.

(Continued)

Table 14.1 (Continued)

Actor	Information	Example Data Sources
	• How are manufacturers doing (Change them? Recall or buy different product from them?)? • Are needs of Operators (customers) being met? • Infrastructure maintenance (with operators) maintenance scheduling to minimize disruption and maximize maintenance.	
Maintainers	• What needs fixing next? • Meeting targets for service levels? • Prediction of failures/maintenance needs. • Manufacturers knowledge of emerging maintenance issues (globally). • New maintenance patterns as a result of changes in timetable?	• Sensors on assets (e.g. on trains, tracks). • Maintenance databases. • Customer relationship databases.
Operators	• What quality level is infrastructure providing? • Is the infrastructure degrading? Does this affect future operations? • What is the demand for operations from users? • Usage demands for period when there is maintenance?	• Customer Feedback. • Driver communication systems. • Sensors in internal sections on an asset. • Timetabling information.
Regulators	• Timetabling. • Competition regulation.	• Timetabling. • Competition regulation.
Users	• Actual service delivery. • Service problems.	• Tweets. • Facebook. • Mobile Apps. • Connected Cars.

a coordinated fashion. In some cases, they are also prevented from doing so; for example, operators are not allowed to share timetabling information, but have to do this via a government-appointed regulatory board.

Some moves in this direction have been made in the guise of the "connected car," where end users' cars are connected via the cloud to a variety of actors that they were previously unconnected to (Ericsson 2012).

Through these sorts of applications, IoT allows drivers and passengers to access applications from a screen within their vehicle that provides them real-time information regarding traffic, congestion, possible parking spots, and tailored navigation services. Sensors in the car will also transmit information to the car manufacturer to improve product development and also provide tailored maintenance schedules for the drivers. City officials can also hook into these data streams in order to gain more detailed understanding of driver behavior, areas of the city that need more planning in order to reduce blackspots, etc.

In order for the transport infrastructure to respond appropriately to the increasing demands, however, the sharing of information from M2M systems needs to be expanded within the city context beyond just cars, and needs to instead provide an integrated transport solution that brings together all the actors involved in a city. Transport is ripe for an IoT solution for several reasons:

- Users often know more about the real-time performance of the transport infrastructure than operators and maintainers, as they are experiencing it. Information can be found using tweets, etc.
- There is a strong link between physical and informational worlds. Sensors on assets, in combination with data from other actors, provide an increased understanding of infrastructure status, maintenance, and product development needs.
- Within the transport sector of many nations, there is already a well-established understanding that these actors need to work together in order to solve extremely complex problems.

Currently, however, the transport sector is held back by the difficulties in sharing information quickly and easily between multiple systems. In the next section we investigate the role of IoT for creating an information marketplace for transport and logistics.

14.6.2 Information marketplace for transport and logistics

In Chapter 3, we outlined a framework for understanding the complexity of information marketplaces within an IoT context. In this section, we apply it directly to transport and logistics (Figure 14.3).

As we outlined in Chapter 3, we will break the IoT solution up into several parts of a value chain: Inputs, Production Processes, and Packaging/Distribution. The technical details of these have been discussed in Chapters 5 through 8, so we do not cover them again here;

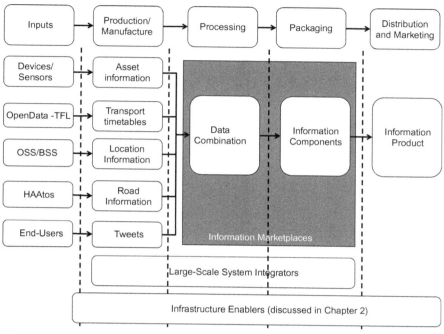

FIGURE 14.3

IoT Information Marketplace for Transport in Cities.

instead we focus on how they connect together to provide a unified information value chain.

Inputs: The devices and sensors here include the sensors on, for example, rail tracks that provide information about the following:

- **Traffic**: Amount and weight of traffic over the road or on a rail track.
- **Subsidence**: Whether the infrastructure in question has moved or changed. These sensors can help indicate when a critical problem on infrastructure may cause a major incident.
- **Infrastructure Temperature and Operating Conditions**: These sensors measure heat, humidity, and other environmental parameters that may affect the asset performance or longevity.
- **Open Data**: Open data may be used as an input into a transport information value chain in the form of maps, transport data, timetable data, and the current performance of the transport network in question.
- **OSS/BSS**: The Operational Support Systems and Business Support Systems of mobile operator networks are also important inputs to

information value chains. Within a transport scenario, they may be used to measure how many people walk between different modes of transport (e.g. between metro and tram or bus). Also, they may be used to understand crowding and congestion in particular areas of a city.

- **Corporate Databases**: A significant number of corporate databases are available within the transport networks of countries. These include the manufacturers' corporate databases of which operators they have sold to, and how the transport asset has performed. Operators in particular have quite large customer relationship management systems that contain a significant amount of detail about customers' preferred travel plans, the performance of their assets, etc.

- **End Users**: Perhaps the most significant input for travel in the new era of IoT is the end user. End users now have the ability to input into the Information Marketplace for transport via, for example, Tweeting about the performance of the transport infrastructure. In addition, a significant majority of smartphones now enable end users to act as their own transport directors; instead of relying solely on the information from operators, they are able to connect together to gain understanding of what their optimal route might be. Moreover, smartphones now contain near field communication (NFC) and other sensor-type technologies that mean that the end-user is able to interact with the transport infrastructure in new, more complex manners. In addition, many end users will be able to pay for their transport via near field technologies in the future. Also, several transport apps allow end users to create "connected journeys," where they are able to create journeys across multiple modes of transport in one city, rather than have to use several websites.

Processing: During the processing stage, data from various sources is mixed together. Making use of the data analytics described in Chapter 5, these data sources are combined together to create insight and information necessary for transport decision-making. Decision-making in the context of transport is tremendously complex. Some examples include:

- Understanding how to best deliver public transport services to lower-income communities.
- Providing appropriate coverage of public transport during peak commuting hours.
- Understanding how to re-route traffic during a major incident.
- Prevention of major rail incidents through deep knowledge of how the transport infrastructure is actually working.

Examples of the processing that are required here include:

- Combining the number of mobile subscribers on a bus at a particular time of day with the timetabling information for local bus services.
- Combining sensor track information owned by the maintenance teams together with the data from train or car manufacturers. This allows a more detailed understanding of how a product is actually performing in use, rather than in the laboratory test environments that manufacturers traditionally have to rely upon.
- Combining detailed demographic information together with maps and environmental sensors that allow city officials to understand how transport routes are affecting the health of different inhabitants in different areas of the city.

Packaging: After the data from various inputs has been combined together, the packaging section of the information value chain creates information components. These components could be produced as charts or other traditional methods of communicating information to end users. Within the transport scenario, the packaging of information would be shared between a broad group of end-user actors. Due to the sensitive nature of the information being shared, it is likely that this information marketplace would be a private one—one that does not make too much data publicly available initially. The actors would therefore be able to package and share the data between one another with an established set of design patterns and data-sharing rules. One difficulty with the packaging of transport information with a city context is the broad number of actors that will need to be able to view and understand the data quickly—everyone from urban planners to transport professionals will need to be able to view and understand the data quickly in order to make decisions. The human interface is therefore one of the most important aspects of this information value chain.

Distribution/Marketing: The final stage of the Information Value Chain is the creation of an Information Product. As we discussed, there are two main categories that these products fall into:

- Information products for improving internal decision-making. These information products are the result of detailed information analysis that allows better decisions to be made. For the transport scenario, cities will be able to create more detailed and streamlined urban planning systems that incorporate IoT data with the vast array of other ICT assets across a city. Information about how a city is actually used by its citizens will give cities the ability to understand how best to deliver transport services while ensuring that the city is a great place to live.

- Information products for resale to other economic actors. These information products have high value for other economic actors and can be sold to them. Within the transport context, while these products may not be directly resold to one another, the ability to share information across the value chain may provide significant value to the economic actors in question. Operators may be able to reduce customer dissatisfaction, and city officials may be able to provide transport services that are enviable to other cities. Infrastructure managers and maintainers, meanwhile, will be able to provide a higher quality service for lower cost while ensuring safety of passengers and employees alike.

14.7 Conclusions

This chapter provided an overview of an example use case of IoT for Smart Cities. The role of IoT in cities is one that will only increase as the pressure on cities to deliver services at reduced costs for an expanding population increases. Many examples exist, including water management, transport management, and waste management. Due to the complexity in cities, the best way for them to achieve the desired outcomes of smart cities is to utilize an information marketplace approach, which allows them to combine together the data from extensive M2M and IoT investments and utilize positive externalities associated with technology in order to reduce environmental impact while creating jobs and economic prosperity for citizens.

Participatory Sensing

15

CHAPTER OUTLINE

15.1 Introduction

Participatory Sensing (PS; Goldman et al., 2009), also known in the literature as Urban, Citizen, or People-Centric Sensing, is a form of citizen engagement for the purposes of capturing the city surrounding environment and daily life. The first PS projects appeared in the early 2000s and focused more on city dweller campaigns to capture problematic situations (such as road faults, air pollution, low-lit parts of the city), daily routines (such as commuting by bike/car), personal individual health, or even combinations thereof. More recently, the concept has gone through a transformation in terms of branding, people engagement practices, target audience, and business model. Nevertheless, the main constituent of any PS activity is the city inhabitants acting as human sensors, and often as end users of an end PS product.

From Machine-to-Machine to the Internet of Things: Introduction to a New Age of Intelligence.
DOI: http://dx.doi.org/10.1016/B978-0-12-407684-6.00015-2

The purpose of early PS was to empower citizens to transform their cities into a collective optimum living environment. With today's technology, citizens have several means of capturing and sharing their daily life and their own view of the quality of living in the city. Most citizens today have mobile phones with which they can capture at least pictures, movies, and sounds, and share them over the Internet. This information can be analyzed and interpreted by individuals, groups, or city officials, and conclusions could be drawn which would result in actions. The information collection can be initiated by a specific sensing/collection campaign organized by any interested actor. This chapter describes the traditional PS concept along with an example case study, as well as the more recent and future trends around the general idea of people sensing. We have chosen the term "Participatory Sensing" for the description of both the early and present concepts for the rest of the chapter.

15.2 Roles, actors, engagement

The main potential actors in a PS system are "individuals" and "city authorities." The city inhabitants have access to sensing devices and can use them to capture their environment. Individuals can be one group of the receivers of the collected information and the target for environmental changes based on PS. The city authorities can be the receiver of the collected information and potentially the organizer of the sensing/collection campaigns. City authorities can also analyze the collected information, potentially with individual citizens, set plans of actions, and follow-up on the actions.

There are several roles for the citizens or city authorities in a PS system. An individual can potentially act as the Data Collector using her mobile phone to collect sensor data. An individual or city authority can play the role of the Collection System Operator who owns and operates the collection system from multiple data collectors. An Analysis Provider that processes, stores, and analyzes the collected data can be assumed by any party (individuals, cities) that could also prepare plans of, and execute actions based on, the conclusions of the campaign (Action Responsible). As a general observation, the city authorities typically assume the non-data collector roles (Collection System Operator, the Analysis Provider, and the Action Responsible) while the citizen's major role is that of a Data Collector, not excluding, however, any other combination of assignments of roles to actors.

Depending on the degree of engagement of the different actors, there are three main models of participation in a PS activity described below.

15.2.1 Collective design and investigation

In this model individual citizens design the sensing campaign, participate in the data collection, and analyze and interpret the collected data. Therefore, the citizens are fully empowered to contribute to the change that they would like to see in their living environment.

15.2.2 Public contribution

Individual citizens only take active participation in the data collection phase organized by another individual or organization (e.g. city authorities), but they don't necessarily analyze or interpret the results.

15.2.3 Personal use and reflection

Individual citizens monitor and record their daily lives for themselves without any organized campaign. A person may choose to refrain from sharing personal information and details, or may choose to share certain specific information or aggregation of collected information. Therefore, the collected data are used mainly for personal reflection or sharing and reflection within a very small and private group (e.g. individual's relatives).

15.3 Participatory sensing process

The basic steps of a typical PS process are shown in Figure 15.1.

During the Coordination phase the participants need to either organize themselves, or be recruited by some other entity (e.g. city authorities) within the context of a sensing campaign, and the objective of the campaign needs to be communicated among all of them. Then the participants

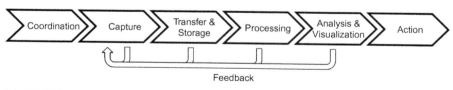

FIGURE 15.1

Typical Participatory Process steps.

spend some predetermined amount of time to capture (Capture phase) the desired sensing modalities using their mobile phone applications or custom designed applications for the sensing campaign. The data entering a PS system does not need to originate only from the data collectors. Several other publicly available sources, such as weather, air quality, and traffic reports, could be used for drawing richer conclusions. The collected data are transferred (Transfer & Storage phase) to the data collection system through the phone connectivity options and stored in Internet servers (private or public). The data are then subject to pre-processing (Process phase) so that the privacy of the data collectors is preserved, and access control rules are added so that the data can be accessed anytime by only authorized individuals or services. The collected data are analyzed by relevant analysis tools, aggregated (if possible), correlated with each other in order to detect patterns, and in the end visualized for better understanding for the target group of the campaign (Analysis and Visualization phase). Last but not least, certain actions (Action phase) may be taken by individuals or city authorities. Feedback is present throughout the whole process and typically assists the Capture phase of the processes. If, for example, the captured data transfer and storage fails, then the participant may be notified to re-transmit or re-capture the target environment. If Processing and Analysis and Visualization result in very little or ambiguous information (e.g. when the processed picture has a bad quality), or participants enter in an area of interest, then they may be notified to (re)capture the situation if possible (Figure 15.1).

15.4 Technology overview

One of the main pieces of technology for a PS campaign is the mobile phone. It encompasses both basic and complex sensors, input and output hardware, one or more processors, one or more modes of communication, location-sensing capabilities, and a software execution environment that can potentially allow the execution of third party software. The minimum capability for a mobile phone in order to be useful for a PS activity is the communication capabilities to allow transmission and reception of messages and potentially sensor data. Active participants can always send text messages with their own sensed data (e.g. pothole on the corner of Main and 3rd Street). The minimum communication capabilities include a cellular (2G, 3G, LTE, WiMAX, etc.), Wi-Fi, or Bluetooth transceiver in order for the participant to potentially receive the campaign start signal (e.g. phone call or short message) and to send the message(s) with captured information. Of course, if for reasons of cost, subscription limitations, or problematic coverage the participant cannot send the messages to the Internet from the cellular transceiver, they can use the short-range transceiver (e.g. Wi-Fi) or Personal Area Network (PAN) transceiver (e.g. Bluetooth) with the cost of a limited geographical reach, and of course, inconvenience. As smart mobile phones begin to dominate the market, inconvenience, and as a result, the barrier for citizen participation, becomes lower and lower. Smart phones are equipped with all kinds of sensors, such as high-resolution photo/video sensors that can measure light intensity, accelerometers, gyroscopes, compasses, location sensors (GPS), infrared sensors, and of course the microphone that could measure a part of the caller's voice and the ambient noise. They are also equipped with some actuators such as displays (e.g. for getting campaign messages), vibrators, or speakers (e.g. for notifying campaign participants that their location is ideal for a capture).

An important requirement for any sensor system, and therefore for a mobile phone as a means for sensing, is that it should have the ability to annotate the collected sensor data with some time and location information. Otherwise the collected data for a single participant cannot be correlated with the other publicly available city sources such as weather reports. Moreover, sensed data missing time or location information cannot be correlated with the corresponding data from other participants, and of course cannot be aggregated to produce useful statistics.

The time and location information is not listed as a minimum requirement on the phone capabilities because the participant can annotate the collected data herself, for example, by writing the time and location in the message that conveys the sensed data; but of course the level of inconvenience is high. Fortunately, location sensors are standard in most modern smart phones, and it is often the case that images and videos are automatically annotated with location information without any intervention from the owner. Time-stamping images and videos are also supported in the majority of modern phones. In the case of other sensor modalities useful for a campaign (e.g. pothole detection by means of a phone accelerometer), a special application is often designed and disseminated as part of the campaign in order to make it convenient for people to participate. For example, there might by an application that always records the sensor data, the time and location, forms the PS message, and dispatches to the data collection system when the time, cost, and network availability conditions are favorable.

Because the sensed data from mobile phones need to be collected in one place for further processing, there is a need for one or more dedicated servers from either a server farm in the premises of the Collection System Operator or from a commercially available Everything as a Service (XaaS) platform (described in Chapter 5). The machines host the appropriate applications for supporting most of the steps in a PS process. While the simplest of the campaigns may need only a simple content database to store sensed data (e.g. ambient noise recordings or photos of the city), more sophisticated campaigns may include pre-processing of each individual data, such as filtering, validation whether or not the collected data is meaningful or not, anonymization of the source of information, removal of people's faces from images/videos, etc. Further actions can also take place such as annotation with city addresses based on location information, compression, storage, etc.

After the necessary data are collected, the Analysis Provider is responsible for analyzing the data. Analysis can be as simple as citizens sifting through the data and discussing with other citizens their conclusions; for example, looking at hundreds of photos of poorly maintained neighborhoods. Or the analysis could be as sophisticated and automatic as determining the light or noise levels of the city according to neighborhoods and time of day, annotating a city map with such information, or creating videos showing the variation of sensed data on a city map. The level of sophistication is clearly up to the imagination of the campaign responsible.

Of course, each PS scenario requires different sensor modalities and different collection processing, analysis, and visualization capabilities. The more organized the campaign in terms of software and hardware capabilities, the lower the barrier for participation and the higher the number of participants.

15.5 An early scenario

In a modern developed city, people move from place to place in order to commute to home, work, school, and extracurricular activities. They walk, drive, or ride private or public means of transportation to get from one point of the city to another, and their mobility is the perfect solution that can ensure as close to complete sensing coverage as possible. The specific use case that we present here involves bikers moving in the city (e.g. commuting between home and work), carrying their mobile phones on themselves (Goldman 2009), and possibly several other sensor devices on their bikes (Figure 15.2)

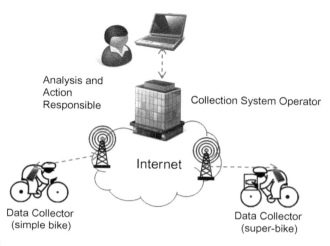

FIGURE 15.2

CycleSense/BikeNet use case.

(Eisenman et al. 2009)

(Eisenman et al. 2009). The bikers can have simple bikes with the only sensing device being their mobile phone, or super-bikes equipped with sensors such as microphones, magnetometers, GPS, CO_2 meters, speedometers, etc. The mobile phone plays the role of the sensor and the communication device. As bikers move in the city, all the data from the sensors are transported to dedicated servers and stored and pre-processed. Individuals or city authorities retrieve anonymized raw data and analyze them as well as let automatic analysis tools produce useful statistics. Raw data are transformed into statistics such as preferred bike routes, traffic problems, road faults, air quality reports, ambient noise levels, and evening light levels, and can be correlated with map and city infrastructure information (e.g. road intersections). Both types of data (raw and analyzed) can be presented to individuals or action-responsible city authorities to remedy problems or contribute to (re) planning of the city infrastructure.

15.6 Recent trends

More recently, the concept of PS has gone through a transformation in terms of branding, people engagement practices, target audience, and business model. The terms used today such as Urban Sensing, Social Sensing, or Citizen Sensing (Sheth 2009, Sheth et al. 2014) don't stress the participatory aspect any more, although people can still participate in sensing campaigns.

The term "campaign" itself implies an active participation for the design and execution of the sensing session, and therefore an active participation to the changes citizen would like to see. However, today's Citizen Sensing focuses more on analyzing and visualizing any data coming from people, regardless of whether people actively design and participate in the on-going research or not. Today's PS activities are mainly uncoordinated, and they are either active or passive. This shift in the engagement practice can be expressed with the introduction of two new participation models, the spontaneous participation or citizen journalism, and the passive or unaware participation.

15.6.1 Citizen journalism

Individuals monitor and record their lives similarly to the personal use and reflection model of early PS models without an organized sensing campaign. This means that they report their findings though social media (e.g. blogs, twitter feeds, social networking web sites), and typically these reports are open to the public, unlike reports that target personal use and reflection. Citizen Journalism is an active sensing engagement, however, the target audience for the sensing campaign is the journalist's own followers/readers/viewers. This type of model is typically used by individuals or authorities during exceptional circumstances such as disasters, without excluding the citizen journalists who regularly post their own version of news. Examples include pictures posted on Facebook about a disaster scene in the city, or tweets providing very terse descriptions of the situation from the individual's point of view. The value of these citizen journalist reports is the freshness of the reports because the individual witnesses are on scene before any city authorities or news correspondents arrive.

15.6.2 Passive participation

The behavior of citizens is captured, stored, and analyzed, with actual citizens often unaware of the fact that their behavior data could be used for the public or the private sector. In such cases, the data are (or should be) anonymized, and possibly aggregated, in order to preserve the privacy of the citizens. Examples include traffic cameras, or electricity metering of certain neighborhoods, or credit card transactions. Either city authorities or private companies collect data about people's behavior and use them for their own purposes, such as city planning or targeted marketing. Therefore, the target audience for the analyzed data is not necessarily citizens or city authorities, but also private sector employees.

The most recent Citizen Sensing activities put more weight on making collection, analysis, and visualization more automated before people can receive the product of the research in order to take action. A human can still be in the loop assisting the automated processes to analyze the collected data better, and of course as an end user of the produced information. The incentive for shifting the focus from the manual processing and analysis to more automated methods is due to the emergence of Big Data Analytics and Semantic Web technologies, both covered in Chapter 5, which promise fast analysis and semantic interpretation of vast amounts of data from both streaming sources (e.g. active or passive citizens) and fixed sources (e.g. cities' own open data). However, these technologies need know-how that volunteers or city authorities don't necessarily have available. Therefore, private companies seize the opportunity to collect data from citizen behavior, analyze them, semantically annotate them, correlate them with other sources, and create information and knowledge. This information and knowledge can be sold in an information marketplace. Examples are the analysis of Twitter messages for detecting market trends or consumer satisfaction.

15.7 A modern example

More recent examples of PS activities focus on exploiting citizen journalist reports from a disaster site in order to produce richer information content and potentially help others who are close to the disaster area. The use case that we describe here consists of three citizen journalists that observe a fire on the corner of Oak and Birch Street near Oak Park (Figure 15.3). At different time instances, each of the participants posts a short message (tweet) with a different description, different intention, and probably spelling and language errors. For example, CityJour#1 names the location of the fire as "Oak Park" CityJour2# as "oak parl," and the third Tweeter as "corner of Oak and Birch street," while there are typos or omissions such as "wins" instead of "winds" As the Twitter feeds allow only short messages, there are typically links to more content for the more interested Tweeter followers/readers; for example, a photo taken from the site annotated with the GPS location. The two Tweeters include such links in their tweet with the links redirecting to another site where photos from the fire are stored.

In terms of the PS process (Figure 15.4), all or most of the spontaneous tweet by the citizen journalists need to be discovered and collected, which is not an easy task. For identifying the related reports analysis on the

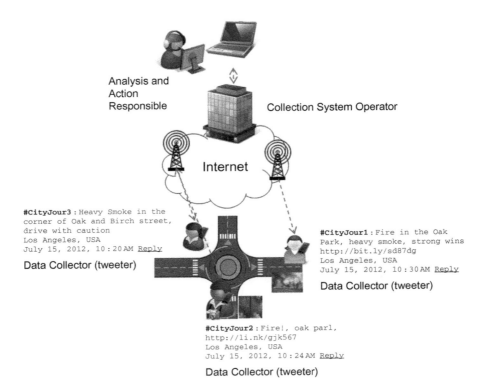

FIGURE 15.3

Citizen Sensing use case.

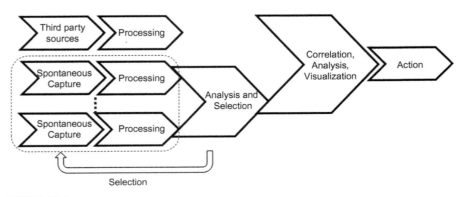

FIGURE 15.4

Citizen Sensing process.

actual text, metadata and linked content at off-site servers may need to take place in order to select the relevant tweets. Moreover, third party, publicly available data sources such as maps or city authority buildings and addresses are used to enable transformation of location information from one format to another (e.g. from GPS coordinates to road names or building addresses). In addition, weather information provides a potential for prediction of the course of the fire. All these sources of information are subject to correlation and analysis before any visualization and action steps are taken. An example of further analysis is semantic analysis of all the sources, extraction of the thematic context, as well as the temporal and spatial context and the fusion of all these pieces of information into one coherent report. In this case the report is that the location of the fire is in Oak Park, specifically on the corner of Oak and Birch Street, and because there are strong winds in the area, the fire is intense and there is heavy smoke that obstructs driving, especially in the intersection. Based on weather reports, the report could be enhanced with fire movement prediction and warnings to drivers and citizens with nearby homes.

Conclusion and Looking Ahead

16

M2M solutions addressing a wide range of different problems and objectives have taken off dramatically over the past few years due to both technology maturity and an expanding market. Wireless sensor network technologies that were in research laboratories 10 years ago are today affordable even at consumer-level costs. In the same way that the PC revolutionized and democratized access to computing, these tools can now be used for real deployment by hobbyists to connect sensors to the Internet and the Web. In addition to this flourishing consumer market, strategic announcements by numerous industrial corporations, large and small, standardization activities in different industry segments, and research communities are coming together to address the needs and to capitalize on the opportunities ahead.

Looking forward, challenges and needs around the IoT will not decrease, but instead grow in importance and magnitude as global issues become increasingly urgent. Handling scarcity of natural resources, reducing impact on climate and environment, and the improving living situations due to increasing urbanization through smart city developments are big issues, but it is now within reach to apply technology to properly manage these issues with the engagement of people and businesses. The focus on solutions will move beyond specific optimizations within a single industry and being entirely business-to-business (B2B)-oriented, to also include consumers and business-to-business-to-consumer (B2B2C). What today is a focus on point problems and isolated business cases is already growing into value creation and innovation, and will increasingly involve and engage people and consumers, enterprises, and society at large. We already see information marketplaces emerging, and that is also one important next step towards IoT.

Technology development will obviously continue to grow at an increasing pace. Embedded technologies and sensing will be further miniaturized and reduced in cost, and we will see an ever-increasing instrumentation of these embedded technologies in the environment around us in the objects and goods we own and use daily. Where technologies have many times

From Machine-to-Machine to the Internet of Things: Introduction to a New Age of Intelligence.
DOI: http://dx.doi.org/10.1016/B978-0-12-407684-6.00016-4

been used for simpler use cases, a shift is already underway with more complex use cases involving heavy machinery, autonomous operations, and advanced remote control of vehicles and robots. In addition, M2M and IoT will create a shift in business models across a number of industries in conjunction with the move from product sales towards a service economy, where a consumer, rather than buying a product, buys a service, where the product is connected for its entire lifetime.

This dual evolution of technology and business drivers will drive horizontal reuse of the deployed embedded technologies, connecting the multitude of small devices and things into the Internet. In combination with the trends towards data and the cloud, data analytics and knowledge-based technologies are major components of the future IoT. Managing the increasing complexity of information and the need to automate actions and control of real-world assets will require new technology developments that go beyond what we today know as Big Data.

For us, the development and evolution of M2M towards the future IoT is just the beginning of a truly connected, smart, and sustainable world.

Abbreviations

3GPP	Third Generation Partnership Project
ADC	Analog-to-Digital Converter
ALG	Application Layer Gateway
API	Application Programming Interface
ARIB	Association of Radio Industries and Businesses
ARIMA	Autoregressive Integrated Moving Average
ARM	Architectural Reference Model
ATIS	Alliance for Telecommunications Industry Solutions
BAN	Body Area Networks
BAS	Building Automation Systems
BBF	Broadband Forum
BOM	Bill of Materials
BSS	Business Support Systems
CAPEX	Capital Expenditures
CCSA	China Communications Standards Association
CENELEC	European Committee for Electrotechnical Standardization
CRISP-DM	Cross Industry Standard Process for Data Mining
CRM	Customer Relationship Management
CWMP	CPE WAN Management Protocol
DA	Device Application
DM	Device Management
DSCL	Device Service Capabilities Layer
DSL	Digital Subscriber Line
EE	Execution Environment
EPCIS	Electronic Product Code Information Services
ERP	Enterprise Resource Planning
ETS	Error-Trend-Seasonal
ETSI	European Telecommunications Standards Institute
GA	Gateway Application
GIS	Geographic Information System
GPIO	General Purpose I/O
GPRS	General Packet Radio Service
GSCL	Gateway Service Capabilities Layer
GSM	Global System for Mobile Communications (originally Groupe Spécial Mobile)
HAN	Home or Building Area Networks
HDFS	Hadoop File System
HTTP	Hypertext Transfer Protocol
HVAC	Heating, Ventilation, and Air Conditioning
ICT	Information and Communications Technology
IERC	Internet of Things European Research Cluster
IETF	Internet Engineering Task Force
IoT	Internet of Things
IoT-A	Internet of Things Architecture
ISO	International Organization for Standardization

ITU	International Telecommunications Union
KDD	Knowledge Discovery in Databases
KMF	Knowledge Management Framework
KPI	Key Performance Indicator
LAN	Local Area Network
LBS	Location Based Services
LTE	Long-Term Evolution
M-BUS	Meter Bus
M2M	Machine-to-Machine
MPP	Massively Parallel Processing
NA	Network Application
NIST	National Institute of Standards and Technology
NSCL	Network Service Capabilities Layer
OMA	Open Mobile Alliance
OPC	OLE for Process Control
OPEX	Operating Expenditure
OS	Operating System
PdM	Predictive Maintenance
PLC	Power Line Communication
QR	Quick Response
RAM	Random Access Memory
RFC	Request For Comments
RFID	Radio Frequency Identification
RPM	Revolutions Per Minute
SaaS	Software as a Service
SAW	Surface Acoustic Wave
SCADA	Supervisory Control and Data Acquisition
SCL	Service Capability Layer
SDK	Software Development Kit
SDO	Standards Developing Organization
SEMMA	Sample, Explore, Modify, Model, and Assess
SoC	System-on-a-Chip
TIA	Telecommunications Industry Association
TTA	Telecommunications Technology Association
TTC	Telecommunication Technology Committee
URL	Uniform Resource Locator
VAV	Variable Air Volume Box
W3C	World Wide Web Consortium
WAN	Wide Area Network
Wi-Fi	Wireless Fidelity
WLAN	Wireless Local Area Network
WPAN	Wireless Personal Area Network

References

ABB. (2013a). <http://new.abb.com/home> Accessed 1.12.13.

ABB. (2013b). *Foundation for the future. Mining magazine*, [online]. Available from <http://www.miningmagazine.com/equipment/foundation-for-the-future> 29.10.13.

ABI. (2012). *Cellular M2M connectivity services*. Available from: ABI Research. [1Q 2012].

ABI Research. (2013). <https://www.abiresearch.com/research/product/1016958-big-data-and-analytics-in-m2m-services/> Accessed 4.12.13.

Azevedo, A., & Santos, M. (2008). KDD, SEMMA and CRISP-DM: A parallel overview. *IADIS European Conference Data Mining*.

BACnet. (2013). BACnet — a data communication protocol for building automation and control networks. <http://www.bacnet.org> Accessed 27.11.13.

Bali, T. G., Demirtas, K. O., & Levy, H. (2009). Is there an intertemporal relation between downside risk and expected returns? *Journal of Financial and Quantitative Analysis*, *44*, 883–909.

Batty, M. (2012). Smart cities, big data. *Environment and Planning B: Planning and Design*, *39*, 191–193.

BCC Research. (2012). Commercial building automation products: technologies and Global markets. Report number: IFT010C.

BDI. (2010). *Internet of energy — ICT for energy markets of the future, federation of German industries (BDI)*. publication No. 439. <http://www.bdi.eu/BDI_english/download_content/ForschungTechnikUndInnovation/BDI_initiative_IoE_us-IdE-Broschure.pdf> Accessed 25.11.13.

Berg. (2013). The global wireless M2M market. Available from: Berg Insight. October 2013.

Bitponics. (2013). <http://www.bitponics.com> Accessed 01.12.13.

Bitponics. (n.d.). Bitponics is your personal gardening assistant. Bitponics. Available from: <http://www.bitponics.com> 29.10.13.

Bizer, C., Heath, T., & Berners-Lee, T. (2009). Linked data - the story so far. *International Journal on Semantic Web and Information Systems (IJSWIS)*, *5*(3), 1–22.

Blue Low Energy Technology. (2013). <https://www.bluetooth.org/en-us/marketing/low-energy-technology> Accessed 27.11.13.

Bluetooth. (2013). Bluetooth special interest group. <http://www.bluetooth.org> Accessed 27.11.13.

Broadband Commission. (2012). The State of Broadband 2012: Achieving digital inclusion for all.

Brand, S. (2006). City planets, Strategy + Business. *Spring*, (42) <http://www.strategy-business.com/article/06109?pg = all> Accessed 23.12.13.

Bryson, J., & Gallagher, P. D. (2012) NIST framework and roadmap for smart grid interoperability standards, release 2.0. Tech. Rep. NIST Special Publication 1108R2, National Institute of Standards and Technology (NIST). <http://www.nist.gov/smartgrid/upload/NIST_Framework_Release_2-0_corr.pdf> Accessed 25.11.13.

Cannata, A., Karnouskos, S., & Taisch, M. (2009). *Dynamic e-maintenance in the era of SOA-ready device dominated industrial environments. IMS — manufacturing technology platform (M4SM), world congress on engineering asset management (WCEAM) conference* Greece: Athens.

Cannata, A., Karnouskos, S., & Taisch, M. (2010). Evaluating the potential of a service oriented infrastructure for the factory of the future. In *8th international conference on industrial informatics (INDIN)*. 13−16.07.2010. Japan: Osaka.

Carrez, F., Bauer, M., Boussard, M., Bui, N., Jardak, C., Loof, J. D., et al. (2013). *Deliverable D1.5 − Final architectural reference model for the IoT v3.0, internet of things − architecture IoT-A EC project deliverable D1.5*. <http://www.iot-a.eu/public/public-documents/d1.5> Accessed 25.11.13.

Castellani, A., Loreto, S., Rahman, A., Fossati, T., & Dijk, E. (2013). *Guidelines for HTTP-CoAP mapping implementations draft-ietf-core-http-mapping-02, IETF CoRE working group*. <http://tools.ietf.org/html/draft-ietf-core-http-mapping-02> Accessed 25.11.13.

Christensen, E., Curbera, F., Meredith, G., & Sanjiva Weerawarana, S. (2001). *Web Services Description Language (WSDL) 1.1, W3C note*. 15 March 2001. <http://www.w3.org/TR/wsdl> Accessed 1.12.13.

CityPulse. (2013). CityPulse: Real-Time IoT stream processing and Large-scale data analytics for smart city applications. Available from: <http://www.ict-citypulse.eu> 19.10.13.

CoBIs (2013). <http://www.cobis-online.de> Accessed 24.11.13.

Colombo, A. W., & Karnouskos, S. (2009). *Towards the factory of the future: A service-oriented cross-layer infrastructure. ICT shaping the world: A scientific view, european telecommunications standards institute (ETSI)*. John Wiley and Sons.

Colombo, A. W., Karnouskos, S., & Mendes, J. M. (2010). Factory of the future: a service-oriented system of modular, dynamic reconfigurable and collaborative systems. In Lyes Benyoucef, & Bernard Grabot (Eds.), *Artificial intelligence techniques for networked manufacturing enterprises management* (p. 2010). Springer.

CONET. (2013). <http://www.cooperating-objects.eu> Accessed 24.11.13.

Davenport, T. H., & Prusak, L. (2000). *Working knowledge: How organizations manage what they know*. Boston, MA: Harvard Business School Press.

De, S., Barnaghi, P., Bauer, M., & Meissner, S. (2011). Service modelling for the internet of things. In: *Proccess Federated Conference on Computer Science and Information Systems (FedCSIS)*, pp. 949−955.

Decker, C., Spiess, P., de Souza, L. M. S., Beigl, M., & Nochta, Z., (2006). Coupling enterprise systems with wireless sensor nodes: Analysis, implementation, experiences and guidelines, Pervasive Technology Applied @ PERVASIVE. Dublin, Ireland.

EC M490 (2011). *Smart GRID mandate, standardization mandate to european standardisation organisations (ESOs) to support european smart grid deployment. M/490 EN*. Brussels: European Commission.

EIF. (2009). *The digital world in 2025, indicators for european action. European internet foundation*. Available from: <http://www.eifonline.org>. 23.09.13.

Eisenman, S. B., Miluzzo, E., Lane, N. D., Peterson, R. A., Ahn, G. -S., & Campbell, A. T. (2009). BikeNet: A mobile sensing system for cyclist experience mapping. *ACM Transactions on Sensor Networks*, 6(1), 87−101.

Ericsson. (2012). More than 50 billion connected devices. *Ericsson White Paper 2012*.

ETSI M2M. (2013). *Machine to machine communications*. <http://www.etsi.org/technologies-clusters/technologies/m2m> Accessed 22.12.13.

ETSI M2M Technical Committee (ETSI M2M TC). (2013a). ETSI TS 102 690 Machine-to-Machine communications (M2M) functional architecture, June 2013. <http://www.etsi.org/deliver/etsi_ts/102600_102699/102690/01.02.01_60/ts_102690v010201p.pdf> Accessed 25.11.13.

ETSI M2M Technical Committee (ETSI M2M TC). (2013b). ETSI TS 102 921 Machine-to-Machine communications (M2M) mIa, dIa and mId interfaces, June 2013. <http://www.etsi.org/deliver/etsi_ts/102900_102999/102921/01.02.01_60/ts_102921v010201p.pdf> Accessed 25.11.13.

ETSI TS 102 689. (2013). Machine-to-Machine communications (M2M); M2M service requirements, August 2010. <http://www.etsi.org/deliver/etsi_ts/102600_102699/102689/01.01.01_60/ts_102689v010101p.pdf> Accessed 25.11.13.

European Commission. (2008). Monitoring and control: Today's market, its evolution till 2020 and the impact of ICT on these. <http://www.decision.eu/smart/SMART_9Oct_v2.pdf> Accessed 24.11.13.

European Commission. (2012). SmartGrids SRA 2035 — strategic research agenda: Update of the smart grids SRA 2007 for the needs by the year 2035, Technology. Report, european technology platform smart grids, european commission. <http://www.smartgrids.eu/documents/sra2035.pdf> Accessed 25.11.13.

Fielding, R. T. (2000). Architectural styles and the design of the network-based software architectures, Ph.D. dissertation, Irvine: University of California.

Food Safety and Modernization Act (FSMA). (January 4, 2011). <http://www.fda.gov/Food/GuidanceRegulation/FSMA/default.htm> Accessed 20.11.13.

FSMA. (2011). FDA Food safety modernization Act. U.S. Food and drug administration. Available from: <http://www.fda.gov/Food/GuidanceRegulation/FSMA/default.htm> 29.10.13.

Furness, A. (2009). Ontology for identification, in CASAGRAS final report, Annex C. <http://www.grifs-project.eu/data/File/Casagras_Final%20Report.pdf> Accessed 1.12.13.

Gereffi, G. (2011). A commodity chains framework for analyzing global industries. *Duke University Working Paper.*

GeoNames (2013). GeoNames ontology <http://www.geonames.org/ontology/documentation.html> Accessed 1.12.13.

Giordano, V., Meletiou, A., Covrig, C. F., Mengolini, A., Ardelean, M., Fulli, G., et al. (2013). Smart grid projects in Europe: Lessons learned and current developments — 2012 update. Joint research center of the european commission, JRC79219, doi:10.2790/82707.

Global Value Chains. (2011). <http://www.globalvaluechains.org> Accessed 25.11.13.

Goldman, J., Shilton, K., Burke, J., Estrin, D., Hansen, M., Ramanathan, N., et al. (2009). Participatory sensing: A citizen-powered approach to illuminating the patterns that shape our world, report on woodrow wilson international center for scholars. <http://www.wilsoncenter.org/publication/participatory-sensing-citizen-powered-approach-to-illuminating-the-patterns-shape-our> Accessed 27.11.13.

GreenWave Reality. (2013). <http://www.GreenWaveReality.com> Accessed 25.11.13.

Gruschka, N., & Gessner, D. (Eds.). (2012). Deliverable 4.2 — concepts and solutions for privacy and security in the resolution infrastructure, internet of things — architecture — project deliverable D4.2. <http://www.iot-a.eu/public/public-documents/documents-1/1/1/d4.2/at_download/file> Accessed 1.12.13.

GS1 EPCGlobal. (2013). GS1 — The gobal language of business. GS1. <HTTP://WWW.GS1.ORG/EPCGLOBAL> Accessed 27.11.13.

Hadley, M. (2009). Web application description language (WADL), W3C member submission, 31 August 2009. <http://www.w3.org/Submission/wadl> Accessed 1.12.13.

Hartke, K. (2013). Observing resources in CoAP draft-ietf-core-observe-11, IETF CoRE working group. <http://tools.ietf.org/html/draft-ietf-core-observe-11> Accessed 25.11.13.

Höglund, J., Ilic, D., Karnouskos, S., Sauter, R., & Goncalves Da Silva, P. (2012). Using a 6LoWPAN smart meter mesh network for event-driven monitoring of power quality. In *Third IEEE international conference on smart grid communications (SmartGridComm)*, Tainan City, Taiwan, 5−8.11.12.

HM Treasury (2011). *National Infrastructure Plan*. November 2011.

IBM. (2008). SOA in manufacturing guidebook, A MESA International, IBM corporation and capgemini co-branded white paper. <ftp://public.dhe.ibm.com/software/plm/pdif/MESA_SOAinManufacturingGuidebook.pdf> Accessed 27.11.13.

IEEE 802.11. (2013). IEEE 802.11 wireless local area networks. Institute of electrical and electronics engineers. <http://www.ieee802.org/11> Accessed 27.11.13.

IEEE 802.15.4. (2013). IEEE 802.15 working group for WPAN. Institute of electrical and electronics engineers. <http://www.ieee802.org/15> Accessed 27.11.13.

IEEE Smart Grid Conceptual Model. (2013). <http://smartgrid.ieee.org/ieee-smart-grid/smart-grid-conceptual-model> Accessed 25.11.13.

IERC. (2013). European research cluster on the internet of things. <http://www.internet-of-things-research.eu> Accessed 27.11.13.

IETF 6LoWPAN BTLE. (2013). Transmission of IPv6 packets over bluetooth low energy, draft-ietf-6lowpan-btle-12, internet engineering task force. <http://tools.ietf.org> Accessed 27.11.13.

Ilic, D., Goncalves Da Silva, P., Karnouskos, S., & Griesemer, M. (2012a). An energy market for trading electricity in smart grid neighbourhoods. In *6th IEEE international conference on digital ecosystem technologies − complex environment engineering (IEEE DEST-CEE)*. Campione d'Italia, Italy, 2012 June.

Ilic, D., Karnouskos, S., & Goncalves Da Silva, P. (2012b). *Sensing in power distribution networks via large numbers of smart meters in the third IEEE PES Innovative Smart Grid Technologies (ISGT) Europe*. Berlin, Germany, 14−17.10.12.

IMC-AESOP. (2013). <http://www.imc-aesop.eu> Accessed 24.11.13.

IMS Research. (2013). Cellular M2M ecosystem - platforms − world. 04.03.13.

Information Marketplaces. The new economics of cities. Available from <http://www.theclimategroup.org/what-we-do/publications/Information-Marketplaces-The-New-Economics-of-Cities/>.

International Telecommunication Union, Telecommunication Standardization Sector (ITU-T). (2013). *ITU-T recommendation Y2060: Overview of internet of things*. Geneva, Switzerland, June 2012. <https://www.itu.int/rec/dologin_pub.asp?lang = e&id = T-REC-Y.2060-201206-I!!PDF-E&type = items> Accessed 25.11.13.

IoT-A (2013) Internet of things architecture. <http://www.ito-a.eu> Accessed 20.10.13.

IoT-A European Commission FP7 Project Consortium. (2013). Requirements − internet of things architecture. <http://www.iot-a.eu/public/requirements/copy_of_requirements> Accessed 11.25.13.

IoT-A UNI (2013). IoT-A unified requirements. <http://www.iot-a.eu/public/requirements> Accessed 26.11.13.

ISA100. (2013). ISA100, Wireless systems for automation, the international society of automation. <http://www.isa.org/isa100> Accessed 27.11.13.

ISO/IEC RFID (2013). Information technology − radio frequency identification for item management. ISO/IEC 18000.

Jain, P., Hedman, P., & Zisimopoulos, H. (2012). Machine type communications in 3GPP systems. *Communications Magazine, IEEE, 50*(11), 28−35. doi:10.1109/MCOM.2012.6353679.

Jasper Wireless. (2013). <http://www.jasperwireless.com> Accessed 25.11.13.

Kadner, K., Oberle, D. (Eds.). (2011). Unified service description language XG final report, W3C incubator group report. 27 October 2011. http://www.w3.org/2005/Incubator/usdl/XGR-usdl/, http://www.internet-of-services.com/index.php?id = 288&L = 0 Accessed 1.12.13.

Karnouskos, S. (2009). Efficient sensor data inclusion in enterprise services. *Datenbank Spektrum Journal, 9*(28), 5−10.

Karnouskos, S. (2011a). Stuxnet worm impact on industrial Cyber-Physical system security, In *37th annual conference of the IEEE industrial electronics society (IECON 2011)*. Melbourne, Australia.

Karnouskos, S. (2011b). Demand side management via prosumer interactions in a smart city energy marketplace, *IEEE International conference on innovative smart grid technologies (ISGT 2011)*, 5−7.12.11. Manchester, UK.

Karnouskos, S., Bangemann, T., & Diedrich, C. (2009a). Integration of legacy devices in the future SOA-based factory. In *13th IFAC symposium on information control problems in manufacturing (INCOM 2009)*, 3−5 June 2009. Moscow, Russia.

Karnouskos, S., & Colombo, A. W. (2011). Architecting the next generation of service-based SCADA/DCS system of systems, in *37th Annual conference of the IEEE industrial electronics society (IECON 2011)*. Melbourne, Australia, 7−10.11.11.

Karnouskos, S., Colombo, A. W., Bangemann, T., Manninen, K., Camp, R., Tilly, M., et al. (2012). A SOA-based architecture for empowering future collaborative cloud-based industrial automation, in *38th annual conference of the IEEE Industrial Electronics Society (IECON 2012)*. Montréal, Canada, 2012 October 25−28.

Karnouskos, S., Goncalves da Silva, P., & Ilic, D. (2011a). Assessment of high-performance smart metering for the web service enabled smart grid, 2nd *ACM/SPEC International Conference on Performance Engineering (ICPE'11)*. Karlsruhe, Germany.

Karnouskos, S., Goncalves Da Silva, P., & Ilic, D. (2013a) Developing a web application for monitoring and management of smart grid neighborhoods. In *IEEE 11th international conference on industrial informatics (INDIN)*. Bochum, Germany, 29−31.07.13.

Karnouskos, S., Guinard, D., Savio, D., Spiess, P., Baecker, O., Trifa, V., et al. (2009b). Towards the Real-Time enterprise: Service-based integration of heterogeneous SOA-ready industrial devices with enterprise applications. In *13th IFAC symposium on information control problems in manufacturing (INCOM)*. Moscow, Russia.

Karnouskos, S., & Haller, S. (2007). Management of hazardous goods with wireless sensor networks. In *International Newsletter on Micro-nano Integration, MSTnews, 5/07*.

Karnouskos, S., Ilic, D., & Goncalves Da Silva, P. (2013b). Assessment of an enterprise energy service platform in a smart grid city pilot. In *IEEE 11th international conference on industrial informatics (INDIN)*. Bochum, Germany, 29−31.07.13.

Karnouskos, S., Savio, D., Spiess, P., Guinard, D., Trifa, V., & Baecker, O. (2010a). Real world service interaction with enterprise systems in dynamic manufacturing environments. In L. Benyoucef, & Bernard Grabot (Eds.), *Artificial intelligence techniques for networked manufacturing enterprises management*. Springer.

Karnouskos, S., & Somlev, V. (2013). Performance assessment of integration in the cloud of things via web services. In *IEEE International Conference on Industrial Technology (ICIT 2013)*. Cape Town, South Africa.

Karnouskos, S., & Terzidis, O. (2007). *Towards an information infrastructure for the future internet of energy. Kommunikation in verteilten systemen (KiVS 2007) conference* Bern, Switzerland: VDE Verlag.

Karnouskos, S., Vilaseñor, V., Handte, M., & Marrón, P. J. (2011b). Ubiquitous integration of cooperating objects. *International Journal of Next-Generation Computing (IJNGC), 2*(3).

Karnouskos, S., Weidlich, A., Kok, K., Warmer, C., Ringelstein, J., Selzam, P., et al. (2010c) Field trials towards integrating smart houses with the smart grid. In *1st International ICST conference on E-Energy, 14−15 October 2010.* Athens Greece.

Karnouskos, S., Weidlich, A., Ringelstein, J., Dimeas, A., Kok, K.Warmer, C., et al. (2010b). Monitoring and control for energy efficiency in the smart house. In *1st international ICST Conference on E-Energy, 14−15 October 2010.* Athens Greece.

KDnuggets. (2007). Polls: Data mining methodology, August 2007. <http://www.kdnuggets.com/polls/2007/data_mining_methodology.htm> Accessed 4.12.13.

KNX. (2013). KNX − the worldwide standard for home and building control. <http://www.knx.org> Accessed 27.11.13.

Kruchten, P. (1995). Architectural blueprints − the '4 + 1' view model of software architecture. *IEEE Software, 12*(6), 42−50.

MakeSense. (2013). <http://www.project-makesense.eu> Accessed 24.11.13.

Marrón, P. J., Karnouskos, S., Minder, D., & Ollero, A. (2011). *The emerging domain of cooperating objects* Springer Verlag.

Martín, G. (Ed.). (2012). *resource description specification,* IoT-A deliverable D2.1. <http://www.iot-a.eu/public/public-documents/documents-1/1/1/copy3_of_d1.2/at_download/file> Accessed 1.12.13.

McKinsey. (2013). Disruptive technologies: Advances that will transform life, business, and the global economy. McKinsey Global Institute. Available from: <http://www.mckinsey.com/insights/business_technoloyg/disruptive_technologies> 10.06.13.

Moore, J. (1996). The death of competition: Leadership and strategy in the age of business ecosystems harperbusiness. ISBN 0-88730-850-3.

Moskovitz, R., & Nikander, P. (2006). Host Identity Protocol (HIP) Architecture, IETF network working group. <http://tools.ietf.org/html/rfc4423> Accessed 2.12.13.

Muller, A., Crespo Marquez, A., & Iung, B. (2008). On the concept of e-maintenance: Review and current research. *Reliability Engineering & System Safety, 93*(8), 1165−1187. doi:10.1016/j.ress.2007.08.006.

Muller, G., Hole, E. (2007). Reference architectures: Why, what and how, white paper architecting forum. <http://www.architectingforum.org/whitepapers/SAF_WhitePaper_2007_4.pdf> Accessed 9.12.13.

Mulligan, C. E. A. (2011). *The communications industries in the era of convergence* Routledge.

NASA JPL. (2011). Semantic Web for Earth and Environmental Terminology (SWEET) ontologies. <http://sweet.jpl.nasa.gov/ontology> Accessed 1.12.13.

NIC. (2012). Global trends 2030: Alternative worlds. National intelligence council. NIC 2013-001. Available from: <http:www.dni.gov/nic/globaltrends> 15.10.13.

NIST. (2011). *The NIST definition of cloud computing.* Special Publication. pp. 800−145.

NOBEL. (2013). <http://www.ict-nobel.eu> Accessed 24.11.13.

Nomagic. (2013). NoMagic UML quick reference guide. <http://www.nomagic.com/images/guides/no_magic_quick_reference_guide_uml.pdf> Accessed 9.12.13.

Object Management Group (OMG). (2013). Unified modelling language. <http://www.uml.org> Accessed 9.1213.

Open Definition. (2013). <http://opendefinition.org> Accessed 25.11.13.

Open Geospatial Consortium (OGC). <http://www.opengeospatial.org/ogc> Accessed 25.11.13.

Open Geospatial Consortium Sensor Web Enablement (OGC SWE) Domain Working Group. (2013). <http://www.opengeospatial.org/projects/groups/sensorwebdwg> Accessed 25.11.13.

Pastor, A., Ho, E., Magerkurth, C., Martín, G., Sáinz, I., Segura, A. S., et al. (2011). Project deliverable D6.2 — updated requirements list, Internet of Things — architecture IoT-A EC FP7 project deliverable D6.2. <http://www.iot-a.eu/public/public-documents/documents-1/1/1/d6.2/at_download/file> Accessed 1.12.13.

Plantagon. (2013). <http://plantagon.com> Accessed 1.12.13.

Plantagon. (n.d.). Feeding the City. *Plantagon*. Available from: <http://www.plantagon.com> 29.10.13.

Rio Tinto. (2012). Mine of the future™, 25 September 2012. <http://www.riotinto.com/documents/120925_JMG_MineExpo.pdf> Accessed 22.11.13.

Rowley, J. (2007). The wisdom hierarchy: Representations of the DIKW hierarchy. *Journal of Information Science*, *33*(2), 163−180.

Rozanski, N., & Woods, E. (2005). Applying viewpoints and views to software architecture. <http://www.viewpoints-and-perspectives.info/vpandp/wp-content/themes/secondedition/doc/VPandV_WhitePaper.pdf>Accessed 2.12.13.

Rozanski, N., & Woods, E. (2011). *Software systems architecture: Working with stakeholders using viewpoints and perspectives* (2nd ed.) Addison-Wesley.

SENSEI. (2013). Integrating the physical with the digital world of the network of the future. <http://www.sensei-project.eu> Accessed 20.11.13).

Shelby, Z. (2012). Constrained RESTful environments (CoRE) link format, IETF CoRE working group, RFC 6690. <http://tools.ietf.org/html/rfc6690> Accessed 25.11.13.

Shelby, Z., Chauvenet, C. (2012). IPSO application framework draft-ipso-app-framework-04, IPSO alliance interop committee. <http://www.ipso-alliance.org/wp-content/media/draft-ipso-app-framework-04.pdf> Accessed 25.11.13.

Shelby, Z., Hartke, K., Bormann, C. (2013a). Constrained Application Protocol (CoAP) draft-ietf-core-coap-18, IETF Draft. <http://tools.ietf.org/html/draft-ietf-core-coap-18> Accessed 25.07.13.

Shelby, Z., Krco, S., & Bormann, C. (2013b). CoRE resource directory draft-ietf-core-resource-directory-00, IETF CoRE working group. <http://tools.ietf.org/html/draft-ietf-core-resource-directory-00> Accessed 9.12.13.

Shelby, Z., Vial, M. V. (2013). CoRE interfaces draft-ietf-core-interfaces-00, IETF CoRE working group. <http://tools.ietf.org/html/draft-ietf-core-interfaces-00> Accessed 25.11.13.

Sheth, A. (2009). Citizen sensing, social signals, and enriching human experience. *IEEE Internet Computing*, *13*(4), 87−92.

Sheth, A., Jadhav, A., Kapanipathi, P., Lu, C., Purohit, H., Smith, G. A., et al. (2014). *Twitris — a system for collective social intelligence, encyclopedia of social network analysis and mining (ESNAM)* Springer.

Shirey, R. (2007). Internet security glossary, Version 2, IETF network working group, RFC4949. <http://tools.ietf.org/html/rfc4949> Accessed 1.12.13.

Singh, S. (2012). *New mega trends — implications for our future lives* London: Palgrave MacMillan.

SmartHouse/SmartGrid. (2013). <http://www.smarthouse-smartgrid.eu> Accessed 24.11.13.

SmartKYE. (2013). <http://www.SmartKYE.eu> Accessed 24.11.13.

SOCRADES. (2013). <http://www.socrades.eu> Accessed 24.11.13.

Software Engineering Institute (SEI), Carnegie Mellon (CMU). (2013). Modern software architecture definitions. <http://www.sei.cmu.edu/architecture/start/glossary/moderndefs.cfm> Accessed 2.12.13.

Sorge, C., Waller, A., Selander, G., Bohli, J. M., Ugus, O., Williams, D. (2010). Deliverable D3.5 - security and accounting for SENSEI. T. Baugé (Ed.), *EC research project SENSEI deliverable D3.5.* [online]. Available at <http://www.ict-sensei.org/index.php?option=com_chronocontact&chronoformname=SENSEI_WP3_D3.5.> Accessed 20.10.13.

Spiess, P., & Karnouskos, S. (2007). Maximizing the business value of networked embedded systems through process-level integration into enterprise software. In *Second International Conference on Pervasive Computing and Applications (ICPCA 2007)*. Birmingham, United Kingdom, pp. 536–541.

Spiess, P., Karnouskos, S., de Souza, L. M. S., Savio, D., Guinard, D., Trifa, V., et al. (2009). Reliable execution of business processes on dynamic networks of service-enabled devices. In *Proc. 7th IEEE International Conference on Industrial Informatics INDIN 2009.* Cardiff, UK, pp. 533–538.

STMicroelectronics. (2013). STM32W-RFCKIT. STM32W Low-cost RF control kit. <http://www.st.com/web/catalog/tools/FM116/SC959/SS1532/PF251361> Accessed 2.12.13.

Sweden Green Building Council. (2013). <http://www.sgbc.se/in-english> Accessed 22.11.13.

Thierauf, R. (1999). *Knowledge management systems for business.* Westport, Conn.: Quorum.

Tinto, R. (n.d.). The mine of the future. Rio Tinto. Available from: <http://www.riotinto.com/ourbusiness/technology-and-innovation-160.aspx> 29.10.13.

Tranquillini, S., Spiess, P., Daniel, F., Karnouskos, S., Casati, F., Oertel, N., et al. (2012). Process-Based design and integration of wireless sensor network applications. *10th international conference on Business Process Management (BPM)*. Tallinn, Estonia.

UN-HABITAT. (2010/2011). State of the World's Cities.

United Nations Statistics Division. (2008). ISIC Rev.4. <http://unstats.un.org/unsd/cr/registry/isic-4.asp> Accessed 23.12.13.

US DOE. (2006). Benefits of demand response in electricity markets and recommendations for achieving them, Tech. rep., US department of energy. <http://energy.gov/sites/prod/files/oeprod/DocumentsandMedia/DOE_Benefits_of_Demand_Response_in_Electricity_Markets_and_Recommendations_for_Achieving_Them_Report_to_Congress.pdf> Accessed 25.11.13.

Vasseur, J. -P., & Dunkels, A. (2010). *Interconnecting smart objects with IP: The next internet* San Francisco, CA, USA: Morgan Kaufmann Publishers Inc.

Vermesan, O., & Friess, P. (Eds.), (2013). *Internet of things: Converging technologies for smart environments and integrated ecosystems* Rover Publishers.

Vial, M. (2012). CoRE mirror server draft-vial-core-mirror-proxy-01, IETF CoRE Working Group. Available at: <https://tools.ietf.org/html/draft-vial-core-mirror-proxy-01> Accessed 25.11.13.

Warmer, C., Kok, K., Karnouskos, S., Weidlich, A., Nestle, D., Selzam, P., et al. (2009). Web services for integration of smart houses in the smart grid. In *Grid-interop - The road to an interoperable grid,* Colorado, USA: Denver, 2009 17–19 November.

Wikimedia Commons. (2013). CRISP-DM Process Diagram. <http://commons.wikimedia.org/wiki/File:CRISP-DM_Process_Diagram.png> Accessed 4.12.13.

ZigBee, I. P. (n.d.). ZigBee IP specification overview. ZigBee alliance. Available from: <http://www.zigbee.org/Specifications/ZigBeeIP/Overview.aspx> 22.10.13.

ZigBee. (SEP 2, n.d.). Smart energy profile 2. ZigBee Alliance. Available from: <http://www.zigbee.org/Standards/ZigBeeSmartEnergy/SmartEnergyProfile2.aspx> 23.10.13.

ZigBee Alliance. (2011). Network device: gateway specification, March 2011. <https://docs.zigbee.org/zigbee-docs/dcn/11/docs-11-5552-00-00mg-zigbee-network-device-zig-bee-gateway-standard-version-1.pdf> Accessed 2.12.13.

ZigBee Alliance. (2013a). Control your world. <http://www.zigbee.org/Specifications.aspx> Accessed 23.12.13.

ZigBee Alliance. (2013b). Smart energy profile (SEP) 2. <http://www.zigbee.org/Standards/ZigBeeSmartEnergy/SmartEnergyProfile2.aspx> Accessed 22.11.13.

ZigBee Alliance. (2013c). ZigBee IP specification overview. <http://www.zigbee.org/Specifications/ZigBeeIP/Overview.aspx> Accessed 22.11.13.

Z-Wave. (2013). Z-Wave: Home control. Sigma designs inc. <http://www.z-wave.com> Accessed 27.11.13.

Index